Lecture Notes in Artificial I

Subseries of Lecture Notes in Compute

Edited by J. G. Carbonell and J. Siekma

Lecture Notes in Computer Science

Edited by G. Goos, J. Hartmanis and J. van Leeuwen

Springer

Berlin
Heidelberg
New York
Barcelona
Budapest
Hong Kong
London
Milan
Paris
Tokyo

Cristiano Castelfranchi
Jean-Pierre Müller (Eds.)

From Reaction
to Cognition

5th European Workshop on Modelling Autonomous
Agents in a Multi-Agent World, MAAMAW '93
Neuchâtel, Switzerland, August 25-27, 1993
Selected Papers

 Springer

Series Editors

Jaime G. Carbonell
School of Computer Science, Carnegie Mellon University
Pittsburgh, PA 15213-3891, USA

Jörg Siekmann
University of Saarland, German Research Center forAI (DFKI)
Stuhlsatzenhausweg 3, D-66123 Saarbrücken, Germany

Volume Editors

Cristiano Castelfranchi
Institute of Psychology, National Research Council — CNR
Viale Marx 15, I-00137 Rome, Italy

Jean-Pierre Müller
Institute of Computer Science andArtificial Intelligence, Neuchâtel University
Rue Emile Argand 11, CH-2007 Neuchâtel, Switzerland

Cataloging-in-Publication Data applied for

Die Deutsche Bibliothek - CIP-Einheitsaufnahme

From reaction to cognition : selected papers / 5th European
Workshop on Modelling Autonomous Agents in a Multi-Agent
World, MAAMAW '93, Neuchâtel, Switzerland, August 25 - 27,
1993. Cristiano Castelfranchi ; Jean-Pierre Müller (ed.). -
Berlin ; Heidelberg ; New York : Springer, 1995
  (Lecture notes in computer science ; Vol. 957 : Lecture notes in
  artificial intelligence)
  ISBN 3-540-60155-4
NE: Castelfranchi, Cristiano [Hrsg.]; European Workshop on Modelling
  Autonomous Agents in a Multi-Agent World <5, 1993, Neuchâtel>;
  GT

CR Subject Classification (1991): I.2, J.4

ISBN 3-540-60155-4 Springer-Verlag Berlin Heidelberg New York

© Springer-Verlag Berlin Heidelberg 1995
Printed in Germany

Typesetting: Camera ready by author
SPIN 10486525   06/3142 – 5 4 3 2 1 0   Printed on acid-free paper

Contents

Multi-Agent Communication

Multi-Agent Architectures

Introduction

From Reaction to Cognition
5th European Workshop on Modelling an Agent
in a Multi-Agent World
MAAMAW '93

Cristiano Castelfranchi[1] and Jean-Pierre Müller[2]

[1] Institute of Psychology, CNR
Roma, Italy
cris@kant.irmkant.rm.cnr.it
[2] Institute of Computer Science and Artificial Intelligence
Neuchâtel University
Neuchâtel, Switzerland
muller@info.unine.ch

1 MAAMAW

The purpose of the European Workshop on Modelling an Agent in a Multi-Agent World (MAAMAW) is to stimulate exchange and discussion of research in the field of multi- agent systems. MAAMAW is the European forum for DAI studies. While classical DAI research was mainly concerned with distributed problem solving and task allocation in view of a common goal, MAAMAW emphasises the problems arising when several autonomous agents, endowed with their own goals, knowledge, and abilities, share a common environment and pursue either shared or competing goals. MAAMAW is therefore interested both in classical DAI problems (co-ordination, communication, co- operation, negotiation, etc.) and in theories of intention and action, or, more generally, in the autonomous agent and multi-agent system architectures. Multi-agent models are a new area of research in rapid growth, of relevant interest both for AI and for social and management sciences.

2 The workshop

In 1993, the workshop was intended to concentrate on one important direction for the future of the Multi-Agent domain, namely the emergence as a new paradigm. The study of "emergence" of functionalities can be used to try bridging the gap between brooks- like, reactive systems (an individual as a set of interacting enti-ties) and cognitive models (an individual as a whole having knowledge, goals and plans), or between implementational and theoretical perspectives, or between the micro and the macro levels of the social action. In particular, the organisations could be seen as entities having knowledge, goals and plans which are not ex-plicitly represented or shared by any individual member but just emerge from the structure and related interactions of the agents composing the organisations. Conversely, part of the knowledge, goals and plans of one agent could be explained by the interaction of the agent with the other agents.

3 The contributions

When looking at the papers collected in this book, this focus have been widely treated from diverse and interesting points of view. Categorising the contributed papers is always hard; first most papers do not let themselves classify in a single category, second the classification probably reflects more the point of view of the editor than of the author. We present our apologises for the latter problem and in order to deal with the former, we will separately present the organisation of the book into five sections and some other way to look at the papers.

As was said before, the theme of emergence has been extensively treated and reveals diversity on the focus of interest regarding emergence in multi-agent systems. Beyond this diversity, the main interest is in the emergence of patterns of interaction which may range from global properties like fairness and efficiency to the formation of social groups passing by strategies and coherent planning behaviours.

3.1 Emergence of global properties

By emergence of global properties, we mean that the pattern of interactions, although easily recognisable or qualifiable by an external observer, has no structural impact on the society of agents but just appears to be there (or not there in certain circumstances).

Alexis Drogoul [7] shows how locally interacting agents can co-ordinate themselves with regular global patterns of behaviours which can be thought from an external observer as strategies. Examples are given for the N-puzzle, the simulation of the sociogenesis of ant colonies and even a distributed chess playing program (where each piece works locally), in each case the behaviour of the system can be described as following identifiable "rules of thumb".

Multi-agent systems certainly have to learn from computer science and especially distributed computing. Regarding the relationship between local and global properties of MAS, Hans-Dieter Burkhard [2] discusses the notion of fairness (the possibility for anyone to proceed). Local fairness (that an agent can proceed when alone), global fairness (at the level of the actions or interactions of the whole group) do not necessarily coincide. In particular, the view an agent has of the system influences the local/global relationship.

In the paper of Katashi Nagao, Koiti Hasida and Takashi Miyata [14], the agent architecture is based on expressing constraints among variables in clausal form to avoid committing to information flow orientation. The degree of violation of each constraint is used as a potential energy which is spread through the network (dynamic computation by generalised back propagation). This activation determines the next action to perform and even a seemingly plan elaboration and execution.

3.2 Emergence of sociality

Under this very broad term, we mean any emergence of groups of agents working together either under (or producing) strong social structures or by simply entering for a while into some dialogue. In these papers, some structuring of the multi-agent system can be observed and not just global properties.

In the paper of Onn Shehory and Sarit Kraus [15], coalitions are produced by decentralised utility maximisation based on information about the utilities and joined utilities of other agents. Utility is explicated in terms of resources and payoffs. If the values of joining coalitions adds, the system converges to the total coalition (everybody is in it). A definition and calculation of Shapley value is given to distribute evenly the coalition extra-payoff among the members.

In the paper of Steven Ketchpel [11], coalitions are produced by decentralised utility maximisation based on information about utilities of other agents, but a distribution of the stable marriage algorithm with unacceptable partners is used. The issues are stability and efficiency. It is shown that Shapley value (explained in the former paper) is fair to divide the coalition utility among the members.

In the two previous papers, coalition (or organisation in the common sense) arises from co-operation. In organisations, the problem is more about the emergence of effective co- operation into existing structures. Natalie S. Glance and Bernardo A. Huberman [10] describe co-operation as an emerging phenomena influenced by the structure of the group (defining mutual influence) and the possibility for individuals to break or join groups called the fluidity. It is shown that flat structures are bound to co- operation defects when the group increases, that fixed hierarchical structures force co- operation but that fluid hierarchical structures are more efficient in maintaining co- operation but with small bursts of defection.

3.3 Multi-agent planning

Most of the previous papers did no or a few assumptions about the mental structures the agents are actually manipulating but the maximisation of some utility function based on their actions and communications. The problem of, at least, manipulating plans is an important topic in the multi-agent domain.

The paper of JyiShane Liu and Katia Sycara [12] bridges the gap between the previous papers and the papers of this section by presenting a multi-agent system from which a global schedule emerges. The multi-agent methodology is applied to job shop scheduling. Constraints are partitioned into types and agents are assigned to a set of constraints of the same types. For example, Job agents to the precedence constraints of each job and resource agents for each resource. The conflict detection is made on conflict of value-assignment to variables (and not on constraint violation) which are the communication medium. A global solution emerges from this interaction (convergence because of a ranking (slack time, etc.)).

K.Fischer, N. Kuhn, H.J.Müller, J.P.Müller and M.Pischel [9] are applying multi-agent systems to transportation problems. Delivery plans are made locally for each truck, co-ordinated at the level of each company (by task allocation mechanisms). Furthermore, the companies negotiate for delivery they can not fulfil alone by exchanging information about free load. The overall behaviour is described in terms of equilibrium with perturbations from internal inference on free load or bad plans and externally from arrival of new orders.

Eitham Ephrati and Jeffrey S. Rosenschein [8] address multi-agent planning by distributing a common goal into individual subtasks, by letting the agents plan locally and then by optimally joining the local plans taking constraints and costs into account in order to determine the best next action of the group. The cycle is repeated until the common goal is achieved, interleaving planning and execution.

3.4 Multi-agent communication

Most of interactions between cognitive agents is made through communication and more specifically speech acts. A serie of papers is devoted to this dimension from a language to specify the communication protocols to semantics and pragmatic defined in terms of mental states.

Focusing on communication, Birgit Burmeister, Afsaneh Haddadi and Kurt Sundermeyer [3] define a protocol on top of message transfer with a syntax of messages, an illocutionary typing for easing interpretation (inform, query, command, etc.), send/receive handlers specification and finally the organisation of these operators into executable networks for specifying protocols (Commanding, Requesting, Proposing,..). This language is embedded into the DASEDIS multi-agent system.

The paper of Milton Correa and Helder Coehlo [correa] presents the pragmatic of communication in MAS based on the mental states of the agent including desires, beliefs, intentions and expectations. The chain of communication is interpreted in terms of mental transformations.

In the section on emergence of sociality, emergence of coalition has been described but not the impact at the individual level. Rosaria Conte and Cristiano Castelfranchi [5] deals with the problem of going from reaction to cognition at the social level: norms are usually seen as emerging from the maximisation of individual utility as long as everybody conforms to the emerging norms. In this paper, the impact of the representation of the norm in each agent and the effect on belief and goal production is described, giving another aspect of the pragmatic of speech acts.

3.5 Multi-agent architectures

Most multi-agent systems concentrate on how to design an agent architecture for dealing with a multi-agent world (partly justifying the title of the european workshops). The papers of this are going from complete agent architectures to more detailed accounts of communication and sociality.

In the paper of P. Bourgine [1], the cognitive architecture of an agent is described as a hierarchy of learning entities controlling other entities and composed of possibly up to three levels: a reactive level providing useful abduction for action/agent selection, a hedonic level learning through reward anticipation and an eductive level using learned sensory-motor anticipation for off-line improvements. The use of Gittins index is advocated for dynamic hierarchies.

Donald Steiner, Alaister Burt, Michael Kolb and Christelle Lerin [16] present the MAI2L concepts which rely on a set of goals which can be planned by using a library of multi-agent plan schemes to be instantiated and elaborated to an executable plan which is further scheduled for execution. A full programming language is provided for plans. The execution can be done by the body (internal execution) or by the communicator through a set of co-operation primitives. Goals are produced internally or in reaction to incoming events/messages. The architecture of an agent is rational, reactive and generic.

Finally, when measuring performance of multi-agent systems, the MxN (M preys, N predators) pursuit problem has very often be used. Mauro Manela and John Campbell [13] use genetic algorithms to produce a multi-agent system and explore the space of possible M and N. Given the fast degradation of performance of multi-agent systems when M and N increase, an optimal (and interesting) class of MxN pursuit problems is proposed together with efficiency measures.

4 Conclusion

The workshop was successful thanks to the high-level contributions and the talks of the three invited speakers:

- Paul Bourgine who talked about emergence of cognition bringing light from artificial life to multi-agent systems[1];
- Katia Sycara who talked about the use of multi-agent systems fro constraint satisfaction problems applied to scheduling (see [12]);
- Michael Georgeff who presented his activities on multi-agent systems using belief-desire-intention architectures.

We cannot conclude without mentioning that important application domains are beginning to be dealt with using multi-agent systems. In this book alone, two application domains are described:

- the job-shop scheduling domain [12]
- the transportation domain [9]

Let us bet that this tendency will increase in the close future.

Cristiano Castelfranchi and Jean-Pierre Müller

Neuchâtel, April 1995

5 Acknowledgement

The chairs would like to thank Giafranca Cerrito, Thouraya Daouas, Khaled Ghedira, Miguel Rodriguez, Miriam Scaglione, François Sprumont and the Institute for Computer Science and Artificial Intelligence of the University of Neuchâtel, for the hard work which contributed to make of this workshop a success. Scientific quality would not have been achieved at this level without the timely and valuable work of the program committee composed of:

Magnus Boman	(Stockholm University and R.I.T. - Sweden)
John Campbell	(University College London - United Kingdom)
Helder Coelho	(INESC, Technical Univ. of Lisbon - Portugal)
Yves Demazeau	(LIFIA/IMAG, Grenoble - France)
Mauro Di Manzo	(Universita di Genova - Italy)
Jean Erceau	(ONERA/GIA, Chatillon - France)
Jacques Ferber	(LAFORIA, Paris - France)
Julia Galliers	(University of Cambridge - United Kingdom)
Hans Haugeneder	(Siemens AG, Muenchen - Germany)
George Kiss	(The Open University, Milton Keynes - United Kingdom)
Paul Levi	(Technischen Universitaet, Muenchen - Germany)
Frank v. Martial	(DETECON, Bonn - Germany)
Maria Miceli	(IP-CNR, Rome - Italy)
Eugenio Oliveira	(Universidade do Porto - Portugal)
John Perram	(Odense Universitet - Denmark)
Jeffrey Rosenschein	(Hebrew University, Jerusalem - Israel)
Walter Van de Velde	(Vrije Universiteit Brussels - Belgium)
Peter Wavish	(Philips Research Lab, Redhill - United Kingdom)
Eric Werner	(INRIA, Sophia-Antipolis - France)
Gilad Zlotkin	(Hebrew University, Jerusalem - Israel)

References

1. P. Bourgine, The Hedonic agent: a constructivist approach of abductive capacities, in this book
2. Hans-Dieter Burkhard, How to define agent properties or: what is a fair agent?, in this book
3. Birgit Burmeister, Afsaneh Haddadi, Kurt Sundermeyer, Generic, Configurable, Cooperation Protocols for Multi-Agent Systems, in this book
4. Cristiano Castelfranchi, Eric Werner, Artificial Social Systems, LNAI 830, Springer-Verlag 1994
5. Rosaria Conte, Cristiano Castelfranchi, Norms as mental objects: from normative beliefs to normative goals, in this book
6. Milton Correa, Helder Coehlo, Around the Architectural Agent Approach to Model Conversations, in this book
7. Alexis Drogoul, When Ants Play Chess, in this book
8. Eitham Ephrati and Jeffrey S. Rosenschein, A Framework for the Interleaving of Execution and Planning for Dynamic Tasks by Multiple Agents, in this book

9. K.Fischer, N.Kuhn, H.J.Müller, J.P.Müller, M.Pischel, Sophisticated and Distributed: The transportation Domain, in this book

10. N.S. Glance, B.A. Huberman, Organizational Fluidity and Sustainable Cooperation, in this book

11. S. Ketchpel, Coalition Formation among Autonomous Agents, in this book

12. JyiShane Liu and Katia Sycara, Emergent Constraint Satisfaction through Multi-Agent Coordinated Interaction, in this book

13. Mauro Manela, John Campbell, Designing Good Pursuit Problems as Tesbeds for Distributed AI: a Novel Application of Genetic Algorithms, in this book

14. Katashi Nagao, Koiti Hasida, Takashi Miyata, Emergent Planning: A Computational Architecture for Situated Behavior, in this book

15. Onn Shehory, Sarit Kraus, Coalition formation among autonomous agents: Strategies and complexity, in this book

16. Donald Steiner, Alaister Burt, Michael Kolb and Christelle Lerin, The Conceptual Framework of MAI2L, in this book

Emergence of global properties

When Ants Play Chess
(Or Can Strategies Emerge From Tactical Behaviours?)

Alexis Drogoul

LAFORIA, Boîte 169, Université Paris VI
4, Place Jussieu 75252 PARIS CEDEX 05
drogoul@laforia.bip.fr

Abstract. Because we think that plans or strategies are useful for co-ordinating multiple agents, and because we hypothesise that most of the plans we use are build partly by us and partly by our immediate environment (which includes other agents), this paper is devoted to the conditions in which strategies can be viewed as the result of interactions between simple agents, each of them having only local information about the state of the world. Our approach is based on the study of some examples of reactive agents applications. Their features are briefly described and we underline, in each of them, what we call the *emergent strategies* obtained from the local interactions between the agents. Three examples are studied this way: the eco-problem-solving implementations of Pengi and the N-Puzzle, and the sociogenesis process occurring in the artificial ant colonies that compose the MANTA project. We then consider a typical strategical game (chess), and see how to decompose it through a distributed reactive approach called MARCH. Some characteristics of the game are analysed and we conclude on the necessity to handle both a global strategy and local tactics in order to obtain a decently strong chess program.

1 Introduction

In military science as well as in everyday life, building a *strategy* usually means making a plan of co-ordinated actions for reaching a particular goal. It often refers to situations in which (1) the goal to reach is possibly conflicting with the goal of another agent and (2) different resources must be employed, sometimes concurrently, in order to reach it. A strategy always relies on two strong assumptions: having a view as global as possible of the current situation, and making sure that the resources will perform the desired actions. That is why military have always paid a great attention to develop intelligence services, in order to obtain as much information as they can, and to establish strong hierarchies, in order to be obeyed on the battlefield. But that is also why famous battles have been lost. The best example could be that of the battle of Waterloo, during which the late arrival of Grouchy, due to both a lack of information and a misunderstanding of orders, has ruined the whole strategy established by Napoleon and its generals. In that respect, the disadvantages of a long term strategy are quite clear (note that these critics can be applied to *plans*, in the AI acceptation of the term, see (Brooks 1987)): (1) relying on too global information prevents it from being truly adaptive to unpredictable details and it is always difficult to backtrack when the resources are already engaged in their respective actions; (2) assuming that the resources are dependable prevents it from being robust to their failures. We can then assume that, without taking external conditions (like weather) into account, the main reasons for which a strategy fails are constituted by local failures at the resources level, be they facing an unexpected event or badly performing their part of the strategy. By analogy with the military terms, we will call these local behaviours *tactical* behaviours.

However, battles are not always lost. And, when they are lost by one of the participants, they are likely to be won by its opponents. In most cases, like in Waterloo from the Allieds' point of view, the success results both from the pre-established strategy and from *tactical* behaviours that were not included in it, but whose occurrence has favoured it. In these cases, the victory can be viewed as the consequence of a set of tactical behaviours, some of them having been generated by a strategy, and some of them not.

But what would an independent observer not aware of both strategies notice ? He would see a sequence of actions, co-ordinated or not, and would reconstruct his own vision of the strategy employed by the winner. In the worst case (see Figure 1), this vision of the battle could be totally irrelevant to the real strategy.

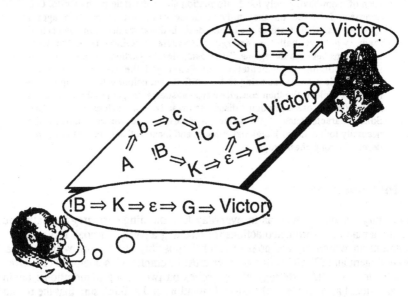

Fig 1 - The strategist (time t-1), the battlefield (time t) and the observer (time t+ 1)

In this somewhat extreme case, the strategist is unuseful and the strategy (or the plan), as understood by the observer, emerges from the co-ordination of local tactical behaviours performed by resources that are not aware of their place in this strategy (for the good reason that it is not written anywhere).
Because we do think that plans or strategies are useful for co-ordinating multiple agents, and because we hypothesise that most of the plans we use everyday are build partly by us, and partly by our immediate environment (which includes other agents),

this paper will be devoted to the conditions in which strategies can be viewed as the result of interactions between simple agents, each of them having only local information about the state of the world. We do not claim that neither all strategies nor all plans can be viewed this way, but we rather think that such an approach can constitute a very constructive lower bound for planning or search, by helping these two fields to take local behaviours into account and by eventually providing them with new tools (for finding local heuristics, for example).

Our approach will be based on the study of some examples of reactive agents applications (see (Ferber & Drogoul 1992) for an exhaustive definition). Their features will be briefly described and we will underline, in each of them, what we call the *emergent strategies* obtained from the local interactions between the agents. Three examples will be studied this way: the eco-problem-solving implementations of Pengi and the N-Puzzle, and the sociogenesis process occurring in the artificial ant colonies that compose the MANTA project.
We will then consider a typical strategical game (chess), and see how to decompose it through a distributed reactive approach called MARCH. Some characteristics of the game will be analysed and we will conclude on the necessity to handle both a global strategy and local tactics in order to obtain a decently strong chess program.

2 Previous works on emergent strategies

2.1 Pengi

Description of the Game

Pengi is based on a video game called Pengo. A penguin (the player) is moving around in a maze made up of ice cubes and diamonds, and inhabited by bees. The penguin must collect all the diamonds while avoiding being stung by a bee or crushed by a cube. It can kill bees by kicking cubes at them. Bees can also kill the penguin by kicking cubes at it or stinging it. They move around in a semi-random way: a bee may sometimes choose to get away from the penguin instead of chasing it. An ice cube can be pushed unless there is another cube adjacent to it in the direction it is pushed. Then it slips along until it hits a cube, a diamond or the border of the screen. Any animal met during its move is crushed.
Therefore, the strategy usually followed by the player is clearly twofold: try to collect as much diamonds as possible while avoiding to be crushed by an ice cube or stung by a bee.

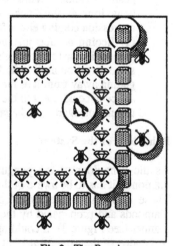

Fig 2 - The Pengi game

The Eco-Problem-Solving Implementation of Pengi

Our implementation of Pengi (different from that of (Agre & Chapman 1987)) has been realised within the Eco-Problem-Solving framework (see (Ferber & Jacopin 1989) or (Ferber & Drogoul 1992) for a definition of it, and (Drogoul & al. 1991) for a more precise description of EcoPengi), in which the agents are provided with a local perception of their environment, a satisfaction behaviour for reaching their own individual goal and a flight behaviour used in case of aggression. All the agents are viewed as totally autonomous, which means that they are not directed by a pre-established strategy. In that respect, their behaviours are:

•   Diamonds are satisfied and cannot flee (i.e. they do not have any behaviours).
•   Ice cubes do not perform any satisfaction behaviour. When being attacked (by a bee or a penguin), they flee in the opposite direction of the aggression and when this location is free, send themselves a flee-message from their old location. During its flight, a cube can enter the perception area of the penguin, and then attack it, or crush an animal when entering its location.
•   Bees own a parametrable perception area (the distance at which they see the penguin). Their satisfaction behaviour is twofold: if the penguin moves around in this area, they take its position as goal and begin to use an incremental satisfaction behaviour (Delaye & al. 1990), which consists in moving each time in the direction of the goal. If not, they randomly choose a goal in this area. They possess two aggressive behaviours: the first one against the ice cubes, the second against the penguin when they see it. They just tell them to flee.
•   The satisfaction behaviour of the penguin consists in reaching and eating the nearest diamond. While not satisfied, it uses the same incremental satisfaction behaviour as the bees. Its flight behaviour is generated by an external aggression coming either from a bee or an ice cube. This behaviour is twofold: (1) fleeing a cube just prompts the penguin to find an adjacent location perpendicular to the cube's moving direction (if there is none, it flees in the same direction until it finds a safe location), and (2) fleeing a bee consists in getting away from it as far as possible (or, if the bee stands on the other side of an ice cube adjacent to the penguin, in kicking it at the bee). Pushing a cube is the unique aggressive behaviour of the penguin.

Behaviour of the System

As the behaviour of the penguin is goal-directed (moving towards the nearest diamond), it has to be aware of the position of the diamonds that it did not eat before. These information are provided to it by means of gradients, generated by the diamonds and propagated by the patches. Although it is computed in a distributed manner (see Figure 3), it could appear as an endeavour to give the penguin a global perception of the game. But what is important to understand is that these gradients do not replace any strategy, because: (1) they provide alternate paths to each diamond and do not force the penguin to follow one of them; (2) they do not provide the penguin with any indication on the co-ordination of actions that it will have to perform to reach a diamond (for example, when meeting bees on the path); (3) they only provide the penguin with the position of the nearest diamond, without paying attention to the other diamonds' positions.

Thanks to this propagation, the penguin just needs, at each time, to ask its patch gradient value, and its incremental satisfaction behaviour will consist in choosing the adjacent location whose gradient value is the smallest.

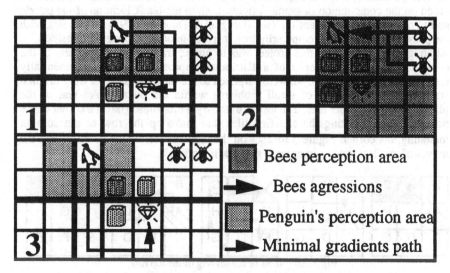

Fig 3 - Two diamonds gradient

When looking at Pengi running, an observer will then face a system that plays (and wins, in most of the cases) the game without defining any *a priori* strategy - which he does not necessarily know. He will see the penguin moving around the board, catching diamonds, avoiding (and sometimes killing) bees and eventually stopping. He will then hypothesise that an intelligent system, hidden somewhere behind the game, has successfully planned the penguin's activity. And he will have this approach because we know that it is difficult to reason about situations like this without a centralised mindset (Resnick 1992). He will then *naturally* build out an *a posteriori* strategy from the sequence of actions of the penguin and attribute it to the omnipotent entity.

Fig 4 - a short-term emergent "plan"

But the long-term strategy as well as the short-term plans (like that on Figure 4) used by the penguin are just artificial reconstructions of the observer's mind, who credits the system with an intelligence that it does not own. For instance, in the situation above, one could think that the penguin has planned a path to the diamond, path that would prevent it from being caught by the bees. But the way the final path has been chosen is only due to the reactivity of the penguin to the bees, and to the alternate paths provided by the gradient propagation.

Pengi allows us to underline three features of the *emergent strategies*: (1) They rarely lead to optimal solutions (the number of moves of the penguin is often greater than the number of moves of a human player); (2) They can be observed but hardly formalised automatically as a set of symbolic representations (see the discussion in (Wavish 1991) about representing emergent behaviours); (3) They are then difficult to use again (which is quite normal, in a way, in a very changing environment).

2.2 The N-Puzzle

Another example of *emergent strategy* can be found in our work on a distributed approach to the N-Puzzle (for a complete description of the solving system see (Drogoul & Dubreuil 1991)). The well known N-puzzle problem consists of a square board containing N square tiles and an empty position called the "blank". Authorised operations slide any tile adjacent to the blank into the blank position. The task is to rearrange the tiles from some random configuration into a particular goal configuration.

Our implementation of the N-Puzzle considers each tile as an autonomous agent, with its own goal and its own perception area. As the world in which these agents move is much more constrained than that of Pengi, the tiles are provided with more sophisticated local heuristics than the penguin. They primarily help them to find a path to their goal without disturbing too much the previously satisfied tiles and without generating loops of aggression. Like Pengi, these behavioural heuristics are based on the computation of gradient fields, both to the blank location of the puzzle and to the goal of the current tile (see (Drogoul & Dubreuil 1993) for details), with the possibility for a tile to temporarily lock its patch in order to protect its position.

Although the first aim of this project was to prove, from the point of view of the computation cost, the interest of a distributed approach in a domain essentially reserved to classical search, we were surprised to see the emergence of truly original strategies in the solving of small problems within the puzzle, like that of the placement of the corner tiles. The task is to satisfy a tile whose goal is a corner of the puzzle without shifting the satisfied tiles that make up the row or the column including this corner. Figure 5 focuses on the problem met by tile A and shows the moves of the fleeing tiles, in order to see how it works.

Fig 5- Snapshots of the solving of the top row

If we comment these stages from the tiles viewpoint, we have, in the beginning, B and C that are satisfied, and A that tries to reach its goal. A then attacks E. The patch of B is provided as a constraint during this attack. E finds no unlocked patch for fleeing. It then unlocks all the patches and attacks A (the patch of B is still a constraint). A flees on the blank. Since it remains unsatisfied, it attacks E again, now with its goal as constraint. E then attacks F, which attacks D, and so on until B flees on the blank. A then slides on its previous patch and attacks B. In this case, there are no valid constraints. B then attacks C, which prefers to slide onto its goal rather than onto the blank. Therefore, it attacks H, which attacks G, which flees onto the blank. A now satisfies itself and the row is solved.

The previous comment describes how the system works, from the point of view of the programmer. It seems quite difficult to understand it as a clear strategy for satisfying the corner tiles. However, the constant repetition of this behaviour during all the experiments gave us the idea to formalise it symbolically, like a naive observer would do, and to try it by hand. And it appeared to be a very powerful and universal "operator" for completing rows and columns, and eventually solving the whole puzzle, even for a human player. The idea is to:

1) Correctly place all the tiles of the row/column except the last one (called Tn), which is quite easy to do.
2) Slide Tn towards its goal without shifting the previous ones.
3) When arrived in front of its goal, slide together Tn and the tile lying on its goal (called Ti) backwards (like A and E on figure 5).
4) Then slide Ti on any patch except the goal and current patch of Tn. This will necessarily shift the previously satisfied tiles.
5) Eventually slide Tn towards its goal while replacing the previous tiles.

Although this algorithm is evidently sub-optimal, it is much simpler than any of the known methods proposed for completing the N-puzzles. It appears to be a good example of the alternate strategies that can be obtained by using a distributed approach to problem solving.

2.3 The Sociogenesis Process in MANTA's colonies

The MANTA project, described in (Drogoul & al. 1991b; Drogoul & Ferber 1992), is an application of the EthoModelling Framework (EMF) to the modelling and simulation of the social organisation in an ant colony. EMF is based on the principles of multi-agent simulation, which means that each organism of the population, ants, cocoons, larvae and eggs, is represented as an agent whose behaviour is programmed with all the required details. Our aim with MANTA is to test hypotheses about the emergence of social structures from the behaviours and interactions of each individual.

The foundation (or sociogenesis) process can be observed in many species of ants. In the species Ectatomma ruidum, whose societies are the natural counterparts of MANTA's simulated ones, the foundation is said semi-claustral : the foundress sometimes leaves her nest to provide the food which is necessary for itself and, above all, for the larvae (in a claustral foundation the queen provides food to the larvae by using her own internal reserves) . Furthermore, the queen generally continues to forage after the emergence of the very first workers.

Sociogenesis is a remarkable challenge at the queen level. As a matter of fact, the queen, which can be considered as an agent intended to behave in a multi-agent system in co-ordination with other agents (the future workers), has to face alone, during the first stages of the society, a very complex situation, in which it has to take care of the whole brood as well as going and find food. So we were interested in the strategy of survival used by the queens, which appear to be quite the same in all natural colonies.

The experiments have been conducted with a model that reproduces the exact laboratory conditions in which natural sociogeneses have been studied. A variable amount of food that depends on the number of ants present in the colony is provided every day (of the simulation time scale) at the entrance of the nest. All the experiments have been tested with the same parameters, and the only source of unpredictability is to be found in the default random walk of the ants.

An experiment begins by putting a queen agent alone in an empty nest and let it evolve (lay eggs, which become larvae, cocoons, and eventually workers). We stop it whenever one of these two situations is reached: (1) the queen dies (by starving); (2) more than six workers are born. The first signifies the death of the colony and the failure of the sociogenesis. The second situation has been chosen as a good estimation of success, with respect to what happens in natural colonies. As a matter of fact, a natural colony that reaches this stage ordinarily keeps on growing, which means that the task of reaching around six workers constitutes the most difficult part of sociogenesis.

Now that we have defined the stop criteria of the simulation, we can take a look at the results. In Table 1 are tabulated the results in terms of successes and failures. Failure cases are clustered in seven categories, each of them indicating the composition of the population when the queen dies.

Results	Composition	Number	%
Failures with	Eggs	4	6,15%
	Eggs, Larvae	10	15,38%
	Larvae	19	29,23%
	Eggs, Larvae, Cocoons	3	4,62%
	Larvae, Cocoons	7	10,77%
	Eggs, Cocoons	1	1,54%
	Larvae, Workers	2	3,08%
Total Number of Failures		46	70,77%
Total Number of Successes		19	29,23%
Total Number of Experiments		65	100,00%

Table 1 - Successes and Failures of the sociogeneses

First point, the proportion of failures (70,77%) appears in these experiments to be close to that observed for natural sociogeneses in laboratory conditions. Second point, the situations in which the foundation of the colony is likely to fail can be obviously characterised by the fact that larvae are part of the population. As a matter of fact, cases of failure in the presence of larvae represent 89,13% of the total number of failures. Why is it so? The simplest explanation that can be provided is, from a behavioural point of view, that larvae must be cared, carried and nourished whereas the other agents composing the brood just need to be cared and carried. The presence of larvae then generates, at the population level, a very important need in food, propagated by the larvae by means of stimuli. Therefore, the agents that can bring back food to the colony (namely the queen and the workers) will have to do it more often, simply because their feeding behaviour will be much more frequently triggered than before. It does not cause too much problems as long as many workers are available, because other tasks still have a good probability to be executed. But, in the early stages of the foundation in which the queen is alone, it quickly results in preventing it from doing anything else, even keeping enough food to feed itself. Moreover, the feeding behaviour is probably the most dangerous behaviour of all, because food is often far away from the brood, which is aggregated near humid places and can then be neglected during a long time, and, more simply, because food can begin to run short, thus obliging the queen to stay outside the nest (or near its entry) until more food is supplied.

However, we have also conducted successful experiments. And, given the constraints stated above, we are going to see which *emergent strategy* is employed by the queen and the first workers to succeed in the foundation of the colony.

The colony whose demographical evolution is depicted on Figure 6 provides us with a typical example of the evolution of the population in our artificial colonies and can advantageously be compared to the curves obtained in the natural sociogeneses. The curve in light grey, which represents the population of eggs, is very unequal, although it is possible to detect a regularity in its sudden falls, which approximately occur every forty days. These falls appear to be synchronised with the peaks of the second curve (in grey), which represents the population of larvae. It is apparent, when looking at other populations, that there is an interplay between the populations of eggs and larvae at the colony level. What is interesting, however, is that this interplay has not been coded in the system. As it could be seen as a particular side effect of a special initial configuration, we have represented, in the six panels of Figure 7, the same curves obtained with six other colonies (from C1 to C6).

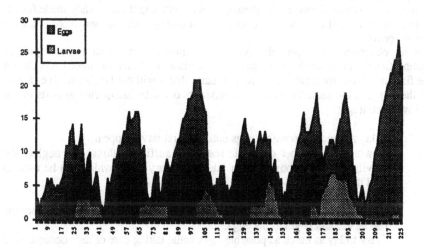

Fig 6 - Populations of eggs and larvae in colony C9

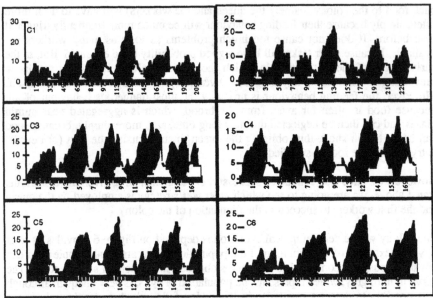

Fig 7 - Populations of eggs and larvae in colonies C1 to C6

All these diagrams clearly show that we obtain the same kind of population fluctuations, with small variations, in different cases. These fluctuations appear to be closely related to those observed in real colonies. As we did not put any global data in the simulations, we can obviously assume that these macro-properties are produced by the behaviours and interactions of the agents at the micro-level. And it is fascinating to see that only a few rules at the individual level can generate such complex patterns at the collective level. From a behavioural point of view, the comments that can be made about these curves are close to those already made for the failure cases, except that the queen succeeds in caring, carrying and feeding all the brood agents.

But an observer could probably credit the queen with much more cognitive capabilities than it is really provided with. As a matter of fact, the queen acts during the first months of foundation as if it anticipates that it will not be able to take care of all the brood at the same time and its behaviours could be interpreted as parts of the following strategy:

1) lay as much as possible eggs until the first larvae appear.
2) when the number of larvae reaches a sufficient threshold, neglect the previous eggs or convert them into alimentary eggs (which can be used to feed the larvae).
3) take care of the larvae until the first cocoons appear.
4) when the number of cocoons reaches a sufficient threshold, neglect the last larvae and begin to lay eggs.
5) try to lay as much as possible eggs while taking care of the cocoons, until the first workers appear.
6) when all the cocoons have become workers, start all over again (but now the workers are able to help the queen in foraging).

A strategy that could be summarised by the following sentence: "never deal with more than two types of brood agents at the same time". And it is very interesting to see that this long-term strategy is simply generated by the interactions between the behaviours of the queen and the needs of the brood agents.

3 A Multi-Agent Reactive Chess Program: MARCH

3.1 Why a Distributed Chess ?

One cannot talk about strategy without talking about chess. Chess appears to be the best example of two-players game in which handling a global strategy is viewed as absolutely necessary for winning. For that reason, we decided to use a distributed approach to implement a chess program, in order to see whether or not local tactical behaviours could really replace a global strategy. Of course, our aim has never been to obtain a chess master. And we did not get one. We simply wanted to program a decent player, able to defend, attack and occasionally mate against average human players, whilst staying as simple as possible. The result, called MARCH (for Multi-Agent Reactive CHess), overpassed much of our expectations on many points (its simplicity/efficiency ratio, its good behaviour in critical situations, etc.), but appeared to be extremely limited on some others (in sacrificing too much pieces, for example), and we believe some of its drawbacks to be due to a lack of certain cognitive capacities. Since MARCH is, as far as we know, the first attempt to apply a multi-agent approach to chess, it is certainly too soon to draw any definitive conclusions from it. But we are convinced that this first steps towards a multi-agent chess program will be useful as a starting point to further works in the domain, which could reconciliate reactive and rational agents.

3.2 Description of MARCH

As for the N-Puzzle, our approach consists in viewing each chessman as an autonomous agent, with its own behaviour and its own perception area. We then have six different types of agents: eight pawns, two knights, two bishops, two rooks, one queen and one king. Each of these agents knows the directions in which it can move, the place it is lying on and its material value (1 for pawns, 3 for knights and bishops, 4 for rooks, 10 for the queen, and infinite for the king).
Each place of the chessboard knows the chessman that is lying on it and holds two values called *whiteStrength* and *blackStrength*.. It also holds the two differences between these values, respectively called *whiteDiff* and *blackDiff*.
A turn consists in:

(1) asking each chess piece to:
> (a) propagate a constant value (namely 10) on the places it directly threatens (generally those on which it could move, except for the pawns), thus increasing the whiteStrength or the blackStrength of the place, depending on the piece's colour.
> (b) inform the enemy's pieces it directly threatens that they are threatened by itself.

(2) asking each threatened pieces to propagate its material value (by adding it to the appropriate strength) on the places situated between itself and the pieces that threaten itself.

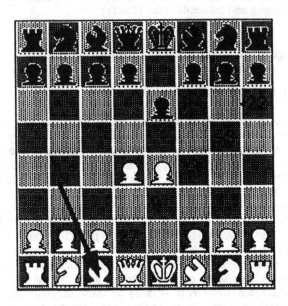

Fig 8- An example of marks given by the white bishop to the places on which it could move

(3) asking each piece to give a mark to each place onto which it could move. This mark is computed as follows:
    (a) it is firstly equal to the whiteDiff or blackDiff of the place, depending on the piece colour.
    (b) if there is any enemy piece on the place, its material value, multiplied by two, is added to the mark.
    (c) the material value of each enemy piece it would threaten when located on this place is added to the mark, as well as the material value of each allied piece it would protect (except that of the king).
    (d) Finally, the material value of the piece and the whiteDiff or black of its place are removed from the mark.
Note that some pieces may mark the places in a different way: the king always rejects places on which the opponent's strength is greater than zero (which means an enemy piece threatens the place); pawns add to the mark a bonus which corresponds to the line number of the place (from 3 or 4 at the beginning to 8 in the enemy lines), because they are inclined to move forwards.

(4) And finally choosing randomly a piece among the pieces whose mark are the greatest and asking it to move on the related place.

An example of marking is shown on Figure 8. The first place gets 27 from the white bishop, which corresponds to its whiteDiff (40 because it is protected by the King, the Queen, the Knight and the bishop), minus the whiteDiff of the bishop's place (10, because it is protected by the queen) and minus the bishop's material value (3). The last place gets -22, which corresponds to its whiteDiff (-10, because it is protected by the bishop but threatened by a black knight and a black pawn), plus the material value of the threatened enemy pieces (1, because there is only a pawn), minus the whiteDiff of the bishop's place (10) and its material value (3).

This system of marking allows each piece to make a careful compromise between security and opportunism, defence and attack. Security is represented by the whiteDiff and blackDiff values, which indicate how much allied pieces are protecting the place and how much enemy pieces are threatening it. Opportunism and attack are represented by the possibilities of capture or threatens a place offers. Defence is represented firstly by whiteDiff or blackDiff, because threatened pieces will artificially increase the strengths of the places between them and their aggressor, thus attracting proportionally to their material value allied pieces that could cover them, and secondly by the possibilities of protecting other pieces.

3.3 Experiments

The experiments with MARCH were firstly conducted by making it play two hundred games against an average human opponent. The program won 57 times, lost 83 times and obtained 60 draws. Almost all its defeats occurred during the first stages of the game and a majority of its victories during the last stages, which means that it is quite bad in opening. However, once its pieces have been correctly deployed, it can play very well and perform clever attacks. Its main drawbacks are a certain predilection towards the systematic exchange of pieces (which partly explains the high number of drawn games), and a very poor defence in "good" situations (like openings).

The second set of experiments was conducted against the GNU Chess program, which is a quite strong player. MARCH did not win any of the fifty games played against it, and obtained only three draws. Once again, almost all its defeats occurred during the first stages of the games.

The preliminary conclusions that we may draw from these experiments are as follows:

(1) We have demonstrated that a multi-agent reactive system can *play* chess with a level approximately equivalent to that of an average human player (i.e. who learned chess in his youth and who does not find the time to train more than once a week...).

(2) Some emergent strategies (sequences of actions that seem to be co-ordinated) can be observed when the system plays (and essentially when it attacks), but they remain partial and do not last during all the game.

(3) When facing a strong strategist (like GNU chess), the program is not able to react in a co-ordinated way and quickly gets trapped.

(4) Obtaining good openings is a very difficult challenge, because the environment of the agents is totally open and not enough threatened by the opponent to make them react intelligently.

4 Conclusion

As we were saying above, MARCH represents the first step towards a strong multi-agent chess program. Our vision is that keeping only reactive agents in it will not allow MARCH to progress significantly. We have certainly explored, with this program, the limits of the emergent strategies that can be obtained from the interactions between reactive agents. Playing chess requires more than a punctual co-ordination between two pieces at a given time.

We do not say that it is not possible to obtain long-term emergent strategies with reactive systems: the example of the sociogenesis proves the contrary. But the difference between the agents that compose an ant colony and the agents that compose the chessboard is that the latter have to face a real strategy whose application is not restricted by any unexpected events or resources failures.

We have shown in this paper that, in many domains, a global strategy could be advantageously replaced by a set of tactical behaviours. This kind of results, although they have to be confirmed by further researches, constitute in our sense an interesting lower bound for studying the influence of individual behaviours on the collective behaviours and the interplays that occur between them.

Bibliography

(Brooks 1987) **R. Brooks**, "Planning is just a way of avoiding figuring out what to do next", MIT Working Paper 303, 1987.

(Delaye & al. 1990) **C. Delaye, J. Ferber & E. Jacopin** "An interactive approach to problem solving"*in* Proceedings of ORSTOM'90, November 1990.

(Drogoul & al. 1991) **A. Drogoul, J. Ferber, E. Jacopin** "Viewing Cognitive Modeling as Eco-Problem-Solving: the Pengi Experience", LAFORIA Technical Report, n°2/91

(Drogoul & Ferber 1992) **A. Drogoul & J. Ferber**, "Multi-Agent Simulation as a Tool for Modeling Societies: Application to Social Differentiation in Ant Colonies", *in* Proceedings of MAAMAW'92 (forthcoming "Decentralized AI IV").

(Drogoul & al. 1992b) **A. Drogoul, J. Ferber, B. Corbara, D. Fresneau,** "A Behavioral Simulation Model for the Study of Emergent Social Structures" *in* Towards a Practive of Autonomous Systems, F.J. Varela & P. Bourgine Eds, pp. 161-170, MIT Press.

(Drogoul & Dubreuil 1992) **A. Drogoul & C. Dubreuil** "Eco-Problem-Solving model: Results of the N-Puzzle", in (Werner & Demazeau 1992), pp 283-295.

(Drogoul & Dubreuil 1993) **A. Drogoul & C. Dubreuil** "A Distributed Approach to N-Puzzle Solving", to appear in the proceedings of the 13th DAI Workshop.

(Ferber & Drogoul 1992) **J. Ferber & A. Drogoul,** "Using Reactive Multi-Agent Systems in Simulation and Problem Solving", in "Distributed Artificial Intelligence: Theory and Praxis", L. Gasser eds.

27

(Wavish 1992) **P. Wavish** "Exploiting Emergent Behaviour in Multi-Agent Systems" *in* (Werner & Demazeau 1992).

(Werner & Demazeau 1992) **E. Werner & Y.Demazeau,** "Decentralized AI 3", North-Holland, June 1992.

How to define agent properties - or: What is a fair agent?

Hans-Dieter Burkhard

Humboldt University Berlin
Dept. of Informatics
Unter den Linden 6, 10099 Berlin, Germany
e-mail: hdb@informatik.hu-berlin.de

Abstract. Different approaches to fairness definitions in multi-agent systems are discussed. They all origin from the classical notion of strong fairness. The underlying problem to be discussed is the faithful description of agent behaviour in a multi-agent system, where the local view clashes with the global dependencies in various ways. The study of fairness properties is one way to examine such problems. Self-determined agents and observation sets are discussed as descriptions of inter-agent dependencies. The relationships between fairness of the system and fairness of the agents are investigated. The results show that fairness analysis needs the analysis of the whole system.

1   Introduction

Studies in multi-agent systems have to deal with properties of the agents and with properties of the whole system. The interactions of both aspects are important concerning the problems of design and of analysis as well. The properties of the agents may depend on their environment. Thereby the global view in a multi-agent system is different from the local view of an agent. It turns out that even the definition of agent properties is a nontrivial problem. In this paper, the study of fairness properties serves as an instance of the overall problems.

Fairness is a property that guarantees each participant of a system eventually to proceed (cf. [10]). In a multi-agent system, this may concern the interactions of the agents, the actions of all agents in the system and the actions of a single agent, respectively. Fairness can be supported by queues etc. which implement a control in some sense. Especially, if fairness is desired for all actions of the system, such control may lead to a global control.

Therefore, the problem arises if fairness properties may be derived from local fairness assumptions. The relationships between global and local properties in multi-agent systems have to be investigated. It should not surprise if global and local properties do not coincide, – related results are known e.g. from theory of games.

In [6] problems of aliveness, deadlock and fairness in multi-agent systems have been considered. It was shown that these different notions lead to different results concerning the relationship between global and local properties. In this paper the (strong) fairness is investigated in more detail. Fairness may be considered as an aspect of cooperative work. Hence studies in fairness can give more insight into cooperative behaviour. But it turns out that even the definition of a "fair agent" can be given in various ways. This leads to closer investigations of what should be called the behaviour of an agent in a multi-agent system.

It is shown that locally defined fairness of the agents can i.g. not be transferred to system fairness. If such a transfer is intended then other global conditions must be fulfilled. These conditions lead to the problem which actions are observable for the agents. It was a hope that the consideration of related "observation sets" would allow some decrease of the global analysis requirements. It is shown that the observation sets allow the recovery of a more realistic system description, but they do not lead to closer relationships between global and local fairness properties.

The paper starts with some basic definitions. Fairness properties are defined as properties of the language describing the behaviour of a system (for simplicity the concurrency is modelled by non-deterministic interleaving).

Multi-agent systems are defined in section 3 in a top-down manner on the base of abstract languages (which again serve as descriptions of the behaviour of the multi-agent system and the agents, respectively). The notion of *self-determined* agents is devoted to a faithful description of the behaviour of the agents.

Different notions of fairness in multi-agent systems are introduced for the system as a whole and for the single agents (section 4). They are shown to be different. For *self-determined* agents some of these differences disappear.

Section 5 gives the results about the relationships between system fairness and fairness of the agents. Some coincidence can be shown only for properties which rely on the whole system (such as *global fairness*, *self-determined* agents, *fair* interaction sets).

The notion of *self-determined* agents is in some cases not satisfactory as long as there is no distinction between "active" and "passive" actions. This leads to the introduction of *observation sets* in section 6 and of a related fairness definition in section 7. But this fairness notion is incomparable to the other ones and hence e.g. global fairness can not be replaced by a less global one under this approach. Nevertheless, the *observation sets* are an interesting approach to study the impact of inter-agent information, communication and observation.

The following notions are used: $\mathbb{N}$ denotes the natural numbers, ∞ denotes "infinitely many". $\forall^\infty$ and $\exists^\infty$ denote "for almost all" and "for infinitely many", respectively.

The set of all finite sequences over a set (alphabet) T is denoted by T^*, e denotes the empty word. The set of all infinite sequences over T is denoted by T^∞.

By $\pi_w(t)$ we denote the number of occurrences of a symbol $t \in T$ in the sequence $w \in T^* \cup T^\infty$ (Parikh-vector).

By $\sqsubset$ and $\sqsubseteq$ we denote the prefix relation. The set of all prefixes of a (finite or infinite) sequence w is denoted by $Pref(w)$. For a set M of sequences the set of all prefixes of these sequences is denoted by $Pref(M)$.

2 Definition of fairness properties

Properties like deadlock avoidance, aliveness and fairness are defined with respect to the behaviour of a system. Thereby the behaviour of a system is built up from atomic actions (or events). These actions can occur sequentially and concurrently. Different calculi have been developed for formalising concurrent behaviour (cf. e.g. [2], [9], [13], [12]), but the simple approach of nondeterministic interleaving and a description using action sequences is sufficient to describe deadlock, aliveness and fairness properties for many purposes. Related approaches are common in the DAI-literature, but mostly they are further exploited using several kinds of logics (e.g. in [14], [8]).

In our approach (cf. [3]) the behaviour of a system can be described by a prefix closed language $L \subseteq T^*$, where T is the finite set of atomic actions of the system. A sequence $p \in L$ describes a possible sequence (history) of actions of the system. Concurrent actions appear in a nondeterministically chosen order. Since each prefix of such a sequence is also a possible behaviour of the system, the language L is prefix closed.

Fairness properties express that each action (or each enabled action) will eventually proceed and not be delayed for ever ([11], [10]). The ultimate condition to eventually proceed is expressed by the notion of impartiality. We obtain the notions of fairness and justice, if the demand to eventually proceed is restricted to those actions which have been enabled.

In each case, fairness properties concern the infinite behaviour of a system. This infinite behaviour is given by the adherence:

Definition 1. Let L be a prefix closed language over a finite set T. The adherence of L is defined by
$$Adh(L) := \{w \in T^\infty \,/\, \forall p \in T^* : p \sqsubset w \to p \in L\}.$$

Example: $Adh(\{a^i b^j / \ i \geq j\}) = \{a^\infty\}$. (The readers are asked for their appreciation of the simple examples. Just as in the case of grammars they point to the principal situation, while more realistic ones would need more space.)

Now *fairness* can be defined in different ways with different meanings. We mention the following ones:

Definition 2. Let L be a prefix closed language over a finite set T.

(a) L is *impartial with respect to* $T' \subseteq T$ (for short: *T'-impartial*)
iff $\forall w \in Adh(L) \forall t \in T' : \pi_w(t) = \infty$.
(b) L is *fair w.r.t.* $T' \subseteq T$ (for short: *T'-fair*)
iff $\forall w \in Adh(L) \forall t \in T' : (\exists^\infty p \sqsubset w : pt \in L) \to \pi_w(t) = \infty$.
(c) L is *just w.r.t.* $T' \subseteq T$ (for short: *T'-just*)
iff $\forall w \in Adh(L) \forall t \in T' : (\forall^\infty p \sqsubset w : pt \in L) \to \pi_w(t) = \infty$.
(d) L is *impartial (fair, just)* iff L is *impartial (fair, just)* w.r.t. T.

Fairness is also called *strong fairness* or *compassion* in the literature [12], while *justice* can be found under the notion of *weak fairness*.

Example: The language $L = \{a^i b^j / i \geq j\}$ is $\{a\}$-*impartial*, $(\{a\}$-*fair*, $\{a\}$-*just*), but not $\{a, b\}$-*impartial* $(\{a, b\}$-*fair*, $\{a, b\}$-*just*).

The following relation hold:

Corollary 3. *Let L be a prefix closed language over a finite set T.*

(a) *If L is* impartial (w.r.t. T') *then L is* fair (w.r.t. T') .

(b) *If L is* fair (w.r.t. T') *then L is* just (w.r.t. T') .

(c) *L is* impartial (w.r.t. T') *iff* $Pref(Adh(L))$ *is* impartial (w.r.t. T').

If there are at least two elements in T then only the implications given or implicated by the corollary hold in general. The language $L = Pref(\{a^{2n} b / n \in \mathbb{N}\})$ is *just* but not *fair*, while $L = \{a^n, b^n / n \in \mathbb{N}\}$ is *fair* but not *impartial*.

By (c) we see that impartiality relies only on the infinite behaviour of a language. All finite languages fulfil the impartiality conditions. But *fairness* and *justice* rely on the whole language (nevertheless finite languages are *fair* and *just*, too).

Our investigations in this paper are restricted to *fairness* only. We mention that *justice* behaves very similar (except e.g. for the *semi-local* notions). *Impartiality* leads to very different results as shown in [6]. This paper gives is a deeper investigation of *fairness* in comparison with the paper [6], thus it may serve as a guide line for further examinations of the other properties in multi-agent systems.

3 Multi-agent systems

Since the properties defined above are definable in terms of languages, it is sufficient for our purposes to consider the behaviour of a system just in the form of a language. Moreover, the further description of a system, e.g. by states and state transitions, is not necessary. Clearly, given an initial state and a sequence of actions, the resulting state can be computed if a transition table is known, but for the consideration of aliveness and fairness properties the states are not obligatory.

In the consequence, it is sufficient to consider and to describe the systems by their behaviour, i .e. a system is given only by some prefix closed language L over a finite alphabet T, where T is the set of atomic actions (or events) and L is the set of all possible action sequences (histories) which could appear in the system. Concurrent behaviour is described by nondeterministic interleaving.

By A we denote a finite set of agents a. Each agent a has a set T_a of its individual actions. This is reflected by the following definition of multi-agent systems [4] .

Definition 4. A multi-agent system (MAS) is given by $M = [A, T, \tau, L]$ where

A is a finite set of agents,

T is a finite set of all actions/events occurring in M,

τ is a mapping from A into the powerset 2^T of T where $T_a := \tau(a)$ is the set of all actions/events from T which are connected with the agent $a \in A$, we suppose $T = \bigcup \{T_a / a \in A\}$,

L is a prefix closed subset of T^* which describes the behaviour of the multi-agent system.

The sets T_a define the restricted knowledge and the restricted influence of the agents a with respect to the whole system. Here we do not suppose that the sets T_a have to be disjoint (which can be desirable from technical reasons in some cases). Thus shared actions or events can be denoted by the same element t (otherwise a shared action/event must be represented by different notations in each set T_a with common occurrences in the sequences from L).

The interpretation of t as an action (an agent is "actively" doing something) or as an event (an agent is more passive, i.e. by observing something) is left open at this time (later on we will make a distinction between "active" and "passive" actions). Thus, if an agent a is doing an action t while b observes this action, this may be described by $t \in T_a$ and $t \in T_b$.

Following these intentions about the sets T_a we define the behaviour of the agents a in a multi-agent systems by the projections of L to the sets T_a:

Definition 5. Let $M = [A, T, \tau, L]$ be a MAS, and $a \in A$. The behaviour of an agent a in M is given by

$$L_a := h_a(L) \ ,$$

where h_a is a homomorphism erasing all $t \notin T_a$ (for $t \in T$: $h_a(t) :=$ if $t \in T_a$ then t else e).

There may be some problems with a faithful definition of the sets T_a. As an example we consider a MAS M with two agents a and b and simply $T_a = \{a\}$ and $T_b = \{b\}$. For the behaviour $L = \{a^i b^j / i \geq j\}$) we have $L_b = \{b\}^*$. The interpretation of this language L_b as the behaviour of the agent b gives the impression that it can perform the action b arbitrarily often. But in the underlying multi-agent system, whenever the agent b starts its work, then there exists an absolute upper bound for the following occurrences of the action b. Such effects must be regarded for a faithful description of fairness properties. It can be done in different ways as the further investigations will show.

One approach to deal with this problem was introduced by the notion of *self-determined* agents:

Definition 6. Let $M = [A, T, \tau, L]$ be a MAS, and $a \in A$. The agent a is *self-determined in M*, iff

$$\forall p, p' \in L : h_a(p) = h_a(p') \rightarrow \forall t \in T_a : (pt \in L \leftrightarrow p't \in L).$$

An agent is *self-determined* if its own information by a sequence $h_a(p)$ about a history $p \in L$ of the system totally determines its next possible actions. This implies that the set T_a has to be large enough to give all the necessary information. In general it must include information about actions of other agents which can be given by "observation actions" or "communication actions" of the agent a, for example.

It can be proved ([5]):

Proposition 7. *Let $M = [A, T, \tau, L]$ be a multi-agent system and let all $a \in A$ be self-determined agents in M. Furthermore let the sets T_a be pairwise disjoint. Then we have $L = Shuff(\{L_a/a \in A\})$.*

Remark. The proposition is only valid if the sets T_a are disjoint.

The conclusion $L = Shuff(\{L_a/a \in A\})$ expresses a total independence of the agents. Vice versa, if the actions of the agents in a MAS are not totally independent of each other, then the preconditions can not hold. In the consequence, a MAS where all agents are *self-determined* but not totally independent from each other, can not be described with disjoint sets T_a.

To become *self-determined* an agent a must have enough information about its environment. For that, its set T_a can be enlarged by introducing "observation actions" which correspond to actions t of other agents b. Proposition (7) shows that it may be necessary to describe these actions by common symbols t in $T_a \cap T_b$. Then new problems may arise for the *self-determination* of agent a if those actions t depend on further actions of other agents. We shall come back to this problem later in section 6.

4 Definition of fairness properties in multi-agent systems

The languages L and L_a (as descriptions of behaviour) are our basis for the fairness definitions.

Definition 8. Let $M = [A, T, \tau, L]$ be a multi-agent system. Then M is *fair* iff L is *fair*.

As already discussed, *fairness* is a property of the infinite behaviour given by the adherence. There are at least three points of view to consider the infinite behaviour of an agent a in a MAS. We can base the definition of *fairness* on

1. the infinite sequences from $Adh(L)$,
2. the infinite sequences from $Adh(L_a) = Adh(h_a(L))$, or on
3. the infinite sequences from $h_a(Adh(L)) \cap T_a^\infty$.

Following these possibilities we define:

Definition 9. Let $M = [A, T, \tau, L]$ be a multi-agent system and let a be an agent from A. The agent a is (a) *globally fair*, (b) *locally fair*, (c) *semi-locally fair* iff

(a) $\forall w \in Adh(L):$ $(\forall t \in T_a : (\exists^\infty q \sqsubseteq w : qt \in L) \to \pi_w(t) = \infty)$,

(b) $\forall w \in (Adh(L_a)):$ $(\forall t \in T_a : (\exists^\infty q \sqsubseteq w : qt \in L_a) \to \pi_w(t) = \infty)$,

(c) $\forall w \in h_a(Adh(L)) \cap T_a^\infty :$ $(\forall t \in T_a : (\exists^\infty q \sqsubseteq w : qt \in L_a) \to \pi_w(t) = \infty)$.

Then it follows immediately by the definitions:

Corollary 10. *Let $M = [A, T, \tau, L]$ be a multi-agent system and let a be an agent from A.*

(a) a *is globally fair iff L is T_a-fair.*
(b) a *is locally fair iff L_a is fair (or equivalently: T_a-fair) .*

The three notions are indeed not equivalent:

Proposition 11. *Let $M = [A, T, \tau, L]$ be a multi-agent system and let a be an agent from A.*

(a) *If a is locally fair then a is semi-locally fair but in general not vice versa.*
(b) *The notions of global fairness and local fairness (semi-local fairness, respectively) are in general not comparable.*

Proof. (a) It follows by Proposition (12a) below that each *locally fair* agent is *semi-locally fair*.

We consider $L = Pref(\{a^n b_1^n b_2 / n \in \mathbb{N}\})$, $T_b = \{b_1, b_2\}$, then b is *semi-locally fair* since $h_b(Adh(L)) \cap T_b^\infty$ is empty but b is not *locally fair* since b_1^∞ is in $Adh(L_b)$ where b_2 is enabled infinitely often.

(b) At first we consider a MAS with $L = Pref(\{b_1^n a b_2 / n \in \mathbb{N}\})$, $T_a = \{a\}$ and $T_b = \{b_1, b_2\}$. Thereby $w = b_1^\infty$ is the only sequence in $Adh(L)$ (and in $h_b(Adh(L))$, too). Only b_1 from T_b is enabled infinitely often during w with respect to L. Hence b is *globally fair*. But since $L_b = Pref(\{b_1^n b_2 / n \in \mathbb{N}\})$, b_2 is enabled infinitely often during w with respect to L_b. Hence b is not *semi-locally fair*. Furthermore, by (a), b is also not *locally fair*. $\square$

Remark. Some other basic notions of *fairness* (instead of "strong fairness") as in [1] would lead to other results.

The "local view" may be misleading, as in the MAS with $L = \{a^i b_1^j b_2^k / i \geq j \geq k\}$ and $T_b = \{b_1, b_2\}$, where $L_b = \{b_1^j b_2^k / j \geq k\}$ is not *fair* such that b is not *locally fair* while in the underlying multi-agent system there does not exist any infinite behaviour for b.

The approach of *semi-local fairness* using the infinite sequences from $h_a(Adh(L)) \cap T^\infty$ seems to be the most adequate notion since on one side it considers only "real" infinite behaviour, but it restricts them to those which are also infinite from the view point of the agent a.

As the examples show, there are different points of view connected with the different definitions, whereby not all views are faithful descriptions of reality. But nevertheless they are interesting with respect to their influence to *fairness* of the whole multi-agent system (cf. the next section).

The differences between the *semi-local* and the *local* point of view are due to differences between $h_a(Adh(L)) \cap T^\infty$ and $Adh(h_a(L))$. We have

$b^{\infty} \in Adh(h_b(\{a^i b^j / i. \geq j\}))$ for $T_b = \{b\}$, but $b^{\infty} \notin h_b(Adh(\{a^i b^j / i \geq j\}))$, whereby the last fact more faithfully expresses reality. It turns out, that the differences disappear for *self-determined* agents:

Proposition 12. *Let* $M = [A, T, \tau, L]$ *be a multi-agent system. Then we have*

(a) $h_a(Adh(L)) \cap T^{\infty} \subseteq Adh(h_a(L))$.

(b) $h_a(Adh(L)) \cap T^{\infty} = Adh(h_a(L))$ *holds if a is a* self-determined *agent.*

Proof. (a) If $w \in T^{\infty}$ is in $h_a(Adh(L))$ then there exists some $w' \in Adh(L)$ with $h_a(w') = w$. Hence for all $u \sqsubset w$ there exists some $u' \sqsubset w'$ with $h_a(u') = u$. Since $w' \in Adh(L)$, all the sequences u' belong to L, and therefore all sequences u belong to $h_a(L)$ such that $w \in Adh(h_a(L))$, which completes the proof.

Note. The inequality may really hold as in the example from above.

(b) To show the equality for the case of *self-determined* agents we make use of a lemma which can be shown by induction over p:

Lemma. *If a is a* self-determined *agent then* $h_a(p) \in L$ *holds for all* $p \in L$.

If $w \in Adh(h_a(L))$ then for all $u \sqsubset w$ we have $u \in h_a(L)$. Furthermore for each such u there exists some $p \in L$ with $h_a(p) = u$. Now we can conclude $h_a(p) = u \in L$ by the lemma. Thus we have obtained that all prefixes u from w belong already to L such that $w \in Adh(L)$. Since w can contain only symbols from T_a, we also have $w \in h_a(Adh(L))$. □

Now, for *self-determined* agents, several distinctions of the fairness notions disappear:

Proposition 13. *Let* $M = [A, T, \tau, L]$ *be a multi-agent system and let a be a* self-determined *agent from A. Then it holds:*

(a) *The notions of* local fairness *and* semi-local fairness *coincide.*

(b) *If a is* globally fair *then a is* (semi-)locally fair, *but in general not vice versa.*

Proof. (a) follows immediately by Proposition (12b).

(b) For $w \in Adh(L_a)$ we have to show that the condition $(\exists^{\infty} q \sqsubset w : qt \in L_a) \rightarrow \pi_w(t) = \infty$ holds if a is *self–determined* and *globally fair*.

First, there exists some $w' \in Adh(L)$ with $h_a(w') = w$ for $w \in Adh(L_a) = Adh(h_a(L))$ by Proposition (12b). Hence for each $q \sqsubset w$ there exists $p \sqsubset w'$ with $h_a(p) = q$. If $qt \in L_a$ then there exists some p' with $p't \in L$. Since again $h_a(p') = q$ and since a is *self–determined*, we obtain $pt \in L$ if $qt \in L_a$.

Now if there are infinitely many prefixes q of w with $qt \in L_a$ then there are infinitely many prefixes p of w' with $pt \in L$. If a is *globally fair* then L is T_a-*fair* and then we must have $\pi_{w'}(t) = \infty$ and hence also $\pi_{h_a(w')}(t) = \pi_w(t) = \infty$ which completes the proof.

By Proposition (11a), *local fairness* implies *semi-local fairness*.

On the other hand, the agent a in the MAS with $L = \{a, b\}^*$, $T_a = \{a\}$ is *self-determined* and *(semi-)locally fair* but not *globally fair*. □

5 Fairness of MAS and fairness of agents

We investigate the correspondence between system fairness and fairness of the agents in this section. In the case of the "local" definitions there is no coincidence. On the other hand, the *global fairness* for the agents coincide with system *fairness*. The following proposition is an immediate consequence of the definitions as long as T is the union of all sets T_a:

Proposition 14. *Let $M = [A, T, \tau, L]$ be a multi-agent system. M is* fair *iff all agents $a \in A$ are globally fair.*

Since *global fairness* is defined (and hence provable) in the context of the whole system, this result may be of limited value for the analysis of systems by analysing the *global fairness* of the agents. The "local" fairness notions may be more relevant. But the next propositions show that the "local" fairness notions do in general not coincide with the system fairness. As already mentioned, the *self-determined* agents are of interest from this reasons as far as for them the *global fairness* implies *(semi-)local fairness*. Via such a correspondence the result from above may be useful.

Proposition 15. *There are multi-agent systems $M = [A, T, \tau, L]$ which are* fair, *but where all agents $a \in A$ are not* (semi-)locally fair.

Proof. The proof is given by a related multi-agent system with agents a and b. We suppose $a_i \in T_a$ and $b_i \in T_b$, respectively. $L = Pref(\{(a_1 b_1)^n b_1 b_2 / n \in \mathbb{N}\} \cup \{(b_1 a_1)^n a_1 a_2 / n \in \mathbb{N}\})$ is *fair*, while a and b are not *semi-locally fair* and hence not *locally fair*. $\square$

Proposition 16. *There are multi-agent systems $M = [A, T, \tau, L]$ which are not* fair, *but where all agents $a \in A$ are* (semi-)locally fair.

Proof. We consider the MAS with $L = \{a, b\}^*$, $T_a = \{a\}$, $T_b = \{b\}$. Thereby the MAS is not *fair*, while the agents are *(semi-)locally fair*. $\square$

Proposition 17. *If a multi-agent system $M = [A, T, \tau, L]$ is* fair *and if all agents $a \in A$ are self-determined, then all agents must be* (semi-)locally fair.

This proposition is a consequence of the Propositions (13b) and (14).

Another relationship between *local fairness* of the agents and *fairness* of the system can be proved using the language of interactions:

Definition 18. *Let $M = [A, T, \tau, L]$ be a multi-agent system where the sets T_a are pairwise disjoint. Then the interactions of the agents of M are given by*

$$L_A := h_A(L) \ ,$$

where h_A is the homomorphism which assigns to each t the owner of this action, i.e., h_A is defined by $h_A(t) := a$ for $t \in T_a$.

Proposition 19. *Let $M = [A, T, \tau, L]$ be a multi-agent system where the sets T_a are pairwise disjoint.*

(a) *If M is fair, then L_A must be fair.*
(b) *If L_A is fair and all agents $a \in A$ are (semi-)locally fair, then M is fair.*

Proof. (a) If there exists $w' \in Adh(L_A)$ with $\exists^\infty q' \sqsubseteq w' : q'a \in L_A$, then there exists $w \in Adh(L)$ with $h_A(w) = w'$ and some $t \in T_a$ with $\exists^\infty q \sqsubseteq w : qt \in L$ (since T_a is finite). Then by *fairness* of L we obtain $\pi_w(t) = \infty$ and hence $\pi_{w'}(a) = \infty$.

(b) For $w \in Adh(L)$ we have to show that the condition $(\exists^\infty q \sqsubseteq w : qt \in L) \to \pi_w(t) = \infty$ holds if L_A is *fair* and if all $a \in A$ are *semi–locally fair*. Corresponding to w there exists $w' = h_A(w) \in Adh(L_A)$, and then $(\exists^\infty q \sqsubseteq w : qt \in L)$ implies $(\exists^\infty q' \sqsubseteq w' : q'a \in L_A)$ for $t \in T_a$. Since L_A is fair, we get $\pi_{w'}(a) = \infty$. Hence, $w'' = h_a(w)$ is infinite, i.e., $w'' \in h_a(Adh(L)) \cap T_a^\infty$. Since all $a \in A$ are *semi–locally fair* we obtain $\pi_w(t) = \pi_{h_a(w)}(t) = \infty$ which completes the proof. □

The *fairness* of L_A is again a global condition of the whole MAS, it can be achieved e.g. by a global control via scheduling the agents.

6 Active and passive actions of an agent

There may occur new problems using the notion of *self-determined* agents. To be *self-determined*, the set T_a of an agent must include enough information about actions/events in the system which are of some relevance for the agent a.

We consider as an example a MAS M with the agents a, b, c and simply $T_a = \{a\}, T_b = \{b\}, T_c = \{c\}$. For the behaviour $L = Pref(\{a^i b^j c^k / i \geq j \geq k\})$ we have $L_c = \{c\}^*$. The interpretation of this language L_c as the behaviour of the agent c gives the impression that it can perform the action c arbitrarily often (even infinitely often since c^∞ is in $Adh(L_c)$). But in the underlying multi-agent system, whenever the agent c starts its work, then there exists an absolute upper bound for the following occurrences of the action c. Obviously, the agent c is not *self-determined*. With the larger set $T_c = \{b, c\}$ and $L_c = \{b^j c^k / j \geq k\}$ the problems concerning the action c disappear, but again we may have problems with the action b, and again the agent c is not *self-determined*. At the end, we could need $T_c = T$ for a "faithful" description.

As already mentioned after Definition (4), the interpretation of t as an action (an agent is "actively" doing something) or as an event (an agent is more "passive", i.e. by observing something) was left open. Thus, if an agent b is doing an action t while c observes this action, this may be described by $t \in T_b$ and $t \in T_c$.

Now this difference becomes important. To know about its own situation the agent c needs to know the actions of b, but it needs not know the exact situation of the agent b (which is determined by the actions of a). This leads to another definition of agents which are "self-determined" in a modified sense while they have enough information about other agents by certain observation actions.

Definition 20. Let $M = [A, T, \tau, L]$ be a MAS, and $a \in A$.
Furthermore, let T'_a be a set with $T_a \subseteq T'_a \subseteq T$, and let h'_a be the homomorphism erasing all $t \notin T'_a$, (i.e. for $t \in T$: $h'_a(t) :=$ if $t \in T'_a$ then t else e).
We call T'_a an *observation set* for a iff
$$\forall p, p' \in L : h'_a(p) = h'_a(p') \rightarrow \forall t \in T_a \ (pt \in L \leftrightarrow p't \in L).$$

Comparing this definition with the definition of *self-determined* agents, we have now demanded that the information by a sequence $h'_a(p)$ (instead of a sequence $h_a(p)$) about the history $p \in L$ is sufficient to determine the next possible actions t from T_a. Thus, T_a as the set of "active" actions may be unchanged, while the set T'_a may be chosen arbitrarily large to be an *observation set*. In this sense the actions from $T'_a - T_a$ can be seen as "passive" actions of the agent a. It can e.g. observe these actions but it has no influence on their occurrence.

Example: $T'_a = \{b, c\}$ is an *observation set* for the agent c in the example from above with $L = \{a^i b^j c^k / i \geq j \geq k\}$.

Corollary 21. *Let* $M = [A, T, \tau, L]$ *be a MAS, and* $a \in A$.

(a) *If* T'_a *is an* observation set *for* a*, then each set* $T''_a \supseteq T'_a$ *is.*
(b) T_a *is an* observation set *iff* a *is self-determined.*
(c) *The set* T *is always an* observation set *for* a*.*

Hence it is always possible to find an appropriate *observation set*, but it may be of interest to find a small one representing a minimum amount of inter-agent information. There may be different minimal *observation sets* for an agent. The *observation sets* may serve to define a minimal amount of communication between the agents in order to provide enough information.

On the other hand, it is not a special property of an agent to have an observation set. This will be the reason that the consideration of fairness properties regarding *observation sets* will not admit a replacement of e.g. *global fairness*.

7   Redefinition of fairness regarding observation sets

The fairness definitions from above are related to the sets T_a of the agents a. In the context of the last section, these actions appear as the active ones for an agent. Setting T_a to be equal to an *observation set* would lead to a *self-determined* agent with the consequences already discussed.

The *(semi-)local fairness* of the agents need not be related to the *fairness* of the system (Propositions (15), (16)). The *global fairness* of the agents is related to the *fairness* of the system (Proposition (14), but it needs the consideration of the whole system for analysing *global fairness* of the agents. In the case of *self-determined* agents, the *local fairness* may be sufficient (Proposition (17)), but thereby again a global property has to be checked since to be *self-determined* is a property which needs the consideration of the whole system. Hence in any case, the analysis of the system can not be split to different analysis steps which regard only (local) properties of the agents.

It could be the case that the analysis of local properties in the context of an *observation set* is sufficient. Thereby the fairness conditions are only demanded for the active actions from T_a. But they can now be considered with respect to an *observation set* T_a', i.e. in the language $h_a'(L)$. Since $T_a \subseteq T_a' \subseteq T$, this is closer to system properties than *local fairness*, but not as close as *global fairness*.

Different definitions of *fairness* could again come from different sets of infinite behaviour as above, but after a closer analysis there remain only two new possibilities. These concern

1. the infinite sequences from $Adh(h_a'(L))$,
2. the infinite sequences from $h_a'(Adh(L)) \cap (T_a')^\infty$.

The fairness conditions for these infinite sequences must hold only with respect to the active actions from T_a. We decide to consider only the first possibility, and give the following definition, where for a moment we do not necessarily suppose to consider an *observation set* T_a'.

Definition 22. Let $M = [A, T, \tau, L]$ be a multi-agent system, let a be an agent from A, and let T_a' be a set with $T_a \subseteq T_a' \subseteq T$.
The agent a is *locally fair in $h_a'(L)$*, iff
$$\forall w \in Adh(h_a'(L)) : (\forall t \in T_a : (\exists^\infty q \sqsubset w : qt \in h_a'(L)) \rightarrow \pi_w(t) = \infty)$$

It follows immediately from the definitions:

Corollary 23. *Let $M = [A, T, \tau, L]$ be a multi-agent system, let a be an agent from A, and let T_a' be an observation set for a. Then it holds:*

(a) *If $T_a' = T$, then a is* locally fair in $h_a'(L)$ *iff a is* globally fair.
(b) *If $T_a' = T_a$, then a is* locally fair in $h_a'(L)$ *iff a is* locally fair.

Now it follows by Proposition (11) that the notion of *local fairness* in $h_a'(L)$ is i.g. incomparable with the notions of *global, semi-local* and *local fairness*, respectively. Moreover, this is even true if T_a' is an observation set. This implies that none of the fairness notions from section 4 can be replaced by our new fairness notion. Especially, the hope that some of the results from earlier sections for *self-determined* agents could be transferred to agents which are *locally fair* in $h_a'(L)$ for an *observation set* T_a' is not realistic.

Proposition 24. *The notion of* local fairness in $h_a'(L)$ *is i.g. incomparable with the notions of* global, semi-local *and* local fairness, *respectively, even if T_a' is an observation set.*

Proof. First we suppose that a is *globally fair in $h_a'(L)$* for an *observation set T_a'*. Since T is always an *observation set*, and since a is *globally fair* in this case, a need not be *(semi-)locally fair* by Proposition (11). If on the other hand $T_a' = T_a$, then a is *locally fair* and it must be *self-determined* (since T_a' is an *observation set*), but even in this case it must not be *globally fair* by Proposition (13).

Now we suppose that a is not *globally fair* in $h'_a(L)$ for an *observation set* T'_a. Again, T is always an *observation set*, and since a is not *globally fair* in this case, a may be *(semi-)locally fair* by Proposition (11).

At last we consider the example $L = \{a^i b^j c^k / i \geq j \geq k\}$ from above where $T'_c = \{b, c\}$ is an *observation set for c*. The agent c is *globally fair*, but not *globally fair* in $h'_c(L) = \{b^j c^k / j \geq k\}$.) $\qquad\qquad\qquad\qquad\square$

8 Conclusions

Several approaches to fairness definitions in multi-agent systems have been discussed. The discussion should be continued. Other basic notions of fairness properties like *impartiality* and *justice* should be investigated, too. As other results in [6] show, the results may be be different from those we have obtained now. All those results may lead to a better understanding of the relationships between agent properties and system properties. Moreover, the faithful modelling of agents and their behaviour needs further investigations. They will lead to a better understanding of the practical problems, too (e.g. what is meant by a "cooperative agent").

The distinction between "active" and "passive" actions by the *observation sets* gives some insight to the need of communication between the agents. It needs further investigation for a better understanding of the global and the local view in a multi-agent system. It could be used also for planning in multi-agent systems.

References

1. Best, E.: Fairness and Conspiracies. *Inform. Process. Lett.* ,18:215–220, 1984.
2. Brookes, S. D., Roscoe, A. W., Winskel, G., editors: *Seminar on Concurrency*. Lect. Notes in Comp. Sci. 197, Springer-Verlag 1985.
3. Burkhard, H. D.: An Investigation of controls for concurrent systems based on abstract control languages. *Theoretical Computer Science*, 38: 193–222, 1985.
4. Burkhard, H. D.: Ein Formalismus für Multi-Agenten-Systeme. *Künstliche Intelligenz*, 6(1): 17–21, March 1992.
5. Burkhard, H. D.: On a formal definition of multi-agent systems. *Workshop Concurrency, Specification & Programming*, Humboldt University Berlin, Nov. 1992.
6. Burkhard, H. D.: Aliveness and Fairness Properties in Multi-Agent Systems. Proc. of the *13th International Joint Conference on Artificial Intelligence*, 325-330, Chambéry, 1993.
7. Gasser, L., Huhns, M. N., editors: *Distributed Artificial Intelligence, Vol. II*. Pitman Publishing/Morgan Kaufmann Publ., San Mateo, CA, 1989.
8. Halpern, J. Y., Moses, Y: Knowledge and Common Knowledge in a Distributed Environment. *J. Assoc. Comput. Mach.* 37(3): 549–587, 1990.
9. Hoare, C. A. R.: *Communicating Sequential Processes*. Prentice Hall, 1985.
10. Kwiatkowska, M.: Survey of fairness notions. *Information and Software Technology* 31(7): 371–386, 1989.

11. Lehmann, D., Pnueli, A., Stavi, J.: Impartiality, justice, fairness: The ethics of concurrent computation. In *Even, S., Kariv, O., (eds.): Automata, Languages and Programming*, Lect. Notes in Comp. Sci. 115, Springer-Verlag, 1981, pp 264–277.
12. Manna, Z., Pnueli, A.: *The Temporal Logic of Reactive and Concurrent Systems*. Springer-Verlag, 1992.
13. Milner, R.: *Communication and Concurrency*. Prentice Hall, 1989.
14. Werner, E.: Cooperating Agents: A Unified Theory of Communication and Social Structure. In [7], pp 3–36.

Emergent Planning: A Computational Architecture for Situated Behaviour

Katashi Nagao[1], Kôiti Hasida[2], and Takashi Miyata[3]

[1] Sony Computer Science Laboratory Inc.,
3-14-13 Higashi-gotanda, Shinagawa-ku, Tokyo 141, Japan
[2] Electrotechnical Laboratory, 1-1-4 Umezono, Tsukuba 305, Japan
[3] University of Tokyo, 7-3-1 Hongo, Bunkyo-ku, Tokyo 113, Japan

Abstract. We discuss planning implemented in a computational architecture called *dynamical constraint programming*. This architecture is totally constraint-based, with the semantics of constraints defined as a sort of dynamics. The degree of violation of constraints is captured in terms of *potential energy*, which is a real-valued function of the state of the constraint. The constraint is thus provided with a fine-grained declarative semantics. Control schemes for analog and symbolic inferences are obtained on the basis of the energy minimisation principle. Information processing occurs as dynamical interaction, so that tight feedback loops are established among diverse sorts of information. Planning is emergent and does not need any specific procedure (i.e., planner). Information processing for action selection (i.e., planning) emerges from the dynamical control of computation. The dynamical state and topology of constraints change in accordance with the interaction between agents and their ever-changing environments. Shifting between reactivity and deliberativity is also an emergent property of this dynamical control.

1 Introduction

It is practically impossible to delimit the information that is potentially relevant to a cognitive agent's behaviour, whereas the behavioural capacity of the cognitive agent is severely restricted. This causes *partiality of information*: the information potentially relevant to the determination of a cognitive agent's action is only partially reflected in its actual behaviour. Due to this, the distribution of relevant information, together with the degree of relevance, drastically varies depending on the context; otherwise a cognitive agent cannot exploit a large enough part of the potentially relevant information across the various situations confronting it.

This gives rise to very diverse patterns of information flow underlying the complex behaviour of the cognitive agent. So cognition is complex, not entirely because the design of the cognitive agent itself is complex, but rather because it is situated in a complex world, which provides the diverse contexts of the cognitive agent's behaviour. The cognitive agent is complex indeed, but still is far simpler than the behaviour of the agent reflecting also the vastness of the world.

To account for this situatedness and relative simplicity of a cognitive agent, the design of a cognitive system should largely abstract away the directions of information flow: the prescribed order of information processing. The models that stipulate the direction of information flow (that is, *procedural* programming) quickly become intractably complex, attributing too much of the complexity of cognitive processes to the complexity of the cognitive system itself, thus failing to capture the situatedness of cognition.

To capture the diversity of information flow, we need *constraint*: a design method without explicit stipulation of domain/task dependent information flow. A system of constraint at least as powerful as first-order logic is considered necessary to design combinatorial behaviours such as language use. However, such powerful formalisms devised so far commit us to untractable computation for maintaining global and crisp consistency. We also need a control scheme for conducting partial and hence tractable computation rather than closure operation, while supporting approximately first-order expressive power and diverse flow of information.

To implement this, we consider a system of constraint represented as a first-order logic program, and postulate a *dynamics* of this constraint. The degree of violation of the constraint is captured in terms of *potential energy*, which is a real-valued function of the state of the constraint. The constraint is thus provided with a fuzzy declarative semantics that is finer-grained than the usual crisp semantics. Control schemes for analog and symbolic inferences are obtained on the basis of the energy minimisation principle.

In this paper, we discuss planning implemented in this architecture called *dynamical constraint programming*. Planning is emergent and requires no mechanism specific to planning (i.e., planner). Information processing for action selection (i.e., planning) emerges from the dynamical control of computation on constraints. The rest of this paper proceeds as follows. In the next section, we discuss interaction between a cognitive agent and its ever-changing environment. Section 3 provides our computational architecture based on constraints with dynamics. Section 4 discusses an example of agent's action planning in a dynamic environment. Section 6 presents the conclusions of the paper.

2   Situated Behaviour as an Emergent Phenomenon

To implement omni-directional information flow in and across a cognitive agent, we employ a constraint-based design method that is orthogonal to information flow. Since a constraint is free from the concept of input/output (information flow between the system and the external world), it does not distinguish between the inside and outside of a system. Constraints are not internal representation of the world, but the world itself, which includes cognitive agents. Seen at the level of information flow, constraints across the boundary between a system and the external world are feedback loops between them. In this view, Brooks [4, 5] and Rosenschein [17] aim to embed an agent (robot) in a dynamic environment by considering the agent's information processing as a part of such feedback loops.

If an environment is fixed, embedding an agent in it is not so difficult. In fact, there are many simple artifacts which are properly embedded in fixed environments, such as vending machines. To be intelligent is not merely to be embedded in an environment but also to flexibly adapt to changes in the world. The ability to adapt and change the structure of feedback loops is hence required. Brooks' robots have more adaptability than Rosenschein's, because Brooks' robots can select relevant feedback loops according to the context. It is unclear, however, how to obtain such adaptivity for 'higher-level' tasks such as language use. Maes [14] proposes how to provide such agents with goals and plans, but expressive power of the underlying representation scheme is restricted, and information flow is highly prescribed because it uses a specific mechanism for planning.

Our design method is free from domain/task dependent prescription of information flow, and the underlying description language has the expressive power of first-order logic, which is enough to deal with complex behaviours such as language use. Specific behaviours such as planning and communication are emergent phenomena in our constraint-based architecture. They emerge from interaction among various constraints under a domain/task-independent dynamical method for controlling information processing.

The boundary between the inside and outside of a system in a feedback loop varies from one situation to another. Information processing is mostly done by the system in some situations, whereas it happens mostly in the environment including the other agents in the other situations. The degree of the system's deliberativity and reactivity is the balance of the amount of information processing between the inside and outside of the system. Since constraint uniformly describes both the inside and the outside of the system, the interactions across the boundary between the system and the external world is accounted for in just the same way with the interactions in the system. Shifting between reactivity and deliberativity is an emergent property of the dynamical control of information processing as well as planning, language use, and so on.

3 Computational Architecture

A quick overview of our computational framework is in order here. See [12, 13] for a more full and general account.

We consider a first-order logic program. This program is treated entirely as a constraint, in the sense that there is no predetermined direction of information flow such as top-down or left-to-right. This means that there is no domain/task-dependent procedures such as a planner. Computation is controlled by *dynamics* in the sense of physics.

A constraint is a set of *clauses*. A clause is basically a disjunction of *literals*. A *literal* is an *atomic constraint* preceded by a sign. An atomic constraint is an *atomic formula* such as p(X,Y,Z)[4] or an *equation* such as X=Y. Signs are '+' and '−'. For any atomic formula α, literal $+\alpha$ stands for just α, and $-\alpha$ stands for

---

[4] A binding is also regarded as an atomic formula. For example, X=f(Y) is an atomic formula with binary predicate =f.

$\neg\alpha$. Names beginning with capital letters represent variables, while other names are predicates. A clause is written as a sequence of included literals followed by a period. The order of literals is not significant. So (1) and (2) represent the same clause, which means that (3) in a rough, crisp approximation.

(1) $-p(U,Y) +q(Z) -U=f(X) -X=Z$.
(2) $+q(Z) -p(f(Z),Y)$.
(3) $\forall U, X, Y \{\neg p(U,Y) \vee q(X) \vee U \neq f(X)\}$

A literal with a null sign is called the *head* of the clause; A clause with a head is called a *definition clause* of the predicate of the head. In a digital approximation, the meaning of this predicate is defined in terms of a necessary and sufficient condition made up from its definition clauses. For instance, if predicate p has the definition clauses in (4), then p is roughly defined by (5).

(4) $p(X) -q(X,a)$.　$p(f(X)) -r(X)$.
(5) $\forall A\{p(A) \Leftrightarrow \{\exists Y(q(A,Y) \wedge Y = a) \vee \exists X(A = f(X) \wedge r(X))\}\}$

Predicates having definition clauses are called *defined predicates*, and the other predicates *free predicates*.

There is only one clause, called the *top clause*, containing literal true. The top clause corresponds to the query in Prolog.[5] That is, top clause (6) represents top-level hypothesis (7).

(6) true $-p(X) +q(X,Y)$.
(7) $\exists X, Y\{p(X) \wedge \neg q(X,Y)\}$

The computation is to tailor the best hypothesis to explain the top-level one. The top clause may change as computation proceeds, in particular when interactions with the world take place.

A constraint is regarded as a network. For instance, clauses (i), (ii), and (iii) are shown graphically in Figure 1. In such a graphical representation, a clause is a closed domain containing the atomic constraints constituting that clause. Atomic constraints without such indication are referred to as negative literals in clauses. An argument of an atomic formula is shown either as a '•' or as an identifier. Equations between arguments are links. Equations in clauses are called *intraclausal equations*, and those outside of clauses are called *extraclausal equations*.

It is important to note that the constraint network contains no objects corresponding to literals apart from atomic formulas. Positive and negative literals are just two types of manifestations of atomic formulas. So for any atomic formula α in the constraint, literals $+\alpha$ and $-\alpha$ both exist even if one of them does not appear in a clause. Also, α may appear in two different clauses as two literals.

We assume that initially all the atomic formulas with the same predicate are unifiable with each other, unless specified otherwise. So at the beginning the

---

[5] Theoretically, Prolog uses false instead of true here. The reason why we use true will be understood on the basis of exclusion energy to be discussed later.

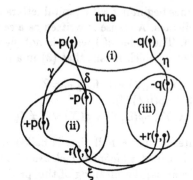

(i) true −p(A) −q(B).
(ii) +p(X) −r(X,Y) −p(Y).
(iii) +r(X,Y) −q(X).

Figure 1: Constraint Network

extraclausal equations constitute a complete graph for every argument place of every predicate.[6] Such a configuration changes as symbolic computation proceeds in the way discussed later.

We postulate a real-valued *potential energy* to measure the degree of violation of the constraint. This defines the declarative semantics. The potential energy is a function of the *activation values* of the atomic constraints (including atomic formulas, bindings, feature specifications and equations) in the constraint. An activation value is a real number varying over time between zero and one. The activation value of an atomic constraint is the degree of its truth, zero corresponding to false and one corresponding to true. The global potential energy is the summation of local potential energies, each of which captures the local semantics of the constraint, such as the disjunction of the literals in a clause, axiom of equation, and so on.

Suppose there are n distinct atomic constraints in the given constraint, and hence n activation values, x_1 through x_n. Then the current state of the system is regarded as a point (8) in the n-dimensional Euclidean space, and the global potential energy E defines a field of force (9).

$$(8) \qquad \vec{x} = \begin{pmatrix} x_1 \\ \vdots \\ x_n \end{pmatrix} \qquad\qquad (9) \quad \vec{F} = -\mathrm{grad}E = \begin{pmatrix} -\frac{\partial E}{\partial x_1} \\ \vdots \\ -\frac{\partial E}{\partial x_n} \end{pmatrix}$$

$\vec{F}$ causes *spreading activation*: when $\vec{F} \neq \vec{0}$, a change of x_i so as to reduce E influences the neighbouring parts of the constraint network, which causes further changes of activation values there, and thus state transition propagates across the network. In the long run, the assignment of the activation values will settle upon a stable equilibrium with $\vec{F} = \vec{0}$. The resulting state gives a minimal value of E. That is, the resultant x satisfies the constraint best in some neighbourhood.

---

[6] There can hence be $O(N^2)$ extraclausal equations, for N different atomic formulas sharing the same predicate. So an efficient encoding schema would be necessary to avoid that space complexity.

The declarative semantics of the entire constraint is decomposed into several aspects. E is the sum of the local energies each representing one such aspect, so that E captures the global declarative semantics. Among the types of energy, here we discuss *disjunction energy, exclusion energy,* and *assimilation energy.*

The *disjunction energy* of a clause captures the disjunction of the literals: at least one literal should be true in a clause. Consider the following clause.

(10) $-p\ +q.$

The disjunctive meaning of this clause is that either $-p$ or $+q$ should be true. The disjunction energy below represents the degree of violation of this meaning.

(11) $Dx_p(1 - x_q)$

D is a positive constant associated with clause (10). Note that (11) is small if and only if either x_p or $(1 - x_q)$ is small; keep in mind that the activation values are between 0 and 1. In dynamic terms, q is excited by p to the extent that p is excited, and p is inhibited by q to the extent that q is inhibited.

Exclusion energy represents the mutual exclusion of the literals in a clause, by which we mean that at most one literal should be true. So in (10), for example, only one of $-p$ and $+q$ may be true. This supports abductive inferences, to assume p when given q, and assume $-q$ when given $-p$. We omit the mathematical details of the energy functions hereafter.

The *assimilation energy* between two unifiable atomic constraints captures the axiom of equality in a relaxed fashion: the activation values of two unifiable atomic constraints should be both close to zero or both close to one to the extent that the extraclausal equations between them are excited. So for instance $p(X,Y)$ and $p(U,V)$ tend to be both excited or both inhibited when the extraclausal equation between X and U and that between Y and V are both strongly excited. The dynamical interaction between two unifiable atomic constraints is stronger when they are in a subsumption relation.

There are two types of computation. Both are controlled by the dynamics derived from the potential energy. One type of computation is spreading activation. Spreading activation attains a locally optimal assignment of activation values to the atomic constraints. The other type is the symbolic operation. Symbolic operations include not only internal operations including *subsumption operation* (a generalisation of unification) and deletion (of links and clauses) but also perception and action. The preference for a symbolic operation is the expected degree $|\frac{\partial U}{\partial w}|$ to which the accompanying local change may dynamically influence *utility function* $U = -P$. P is the energy of the top level of the constraint at the equilibrium of spreading activation. So U is a function of parameters of the energy function. w is some parameter of the potential energy that is altered by the operation in question. When U is large, it means that the system is subjectively satisfied. $|\frac{\partial U}{\partial w}|$ is regarded as the importance of parameter w and hence the part of the constraint to which w is assigned. The preference for an action is given by $\frac{\partial U}{\partial e}$. e is the external input to the atomic formula corresponding to that action. This external input is dummy. That is, e is always 0. Values $\frac{\partial U}{\partial w}$ and $\frac{\partial U}{\partial e}$ are computed by *generalised backpropagation* [16].

This architecture is omni-directional and holistic for the following reasons. First, the spreading activation based on potential energy implements isotropic information flow, because any two activation values influence each other to the extent that they are related. Second, in this connection, backpropagation is also bidirectional in the same sense, the importance of any two parameters affecting each other to the extent that those parameters are related. Third, symbolic operations do not follow any prescribed order or direction, but are controlled on the basis of such dynamical properties. Each symbolic operation is affected by the local dynamical context, which reflects the global context due to generalised backpropagation.

4 Example

In this section, we consider the planning process of cognitive agents s (the speaker) and h (the hearer) in a situation where s wants salt. s first tries to take salt, but fails to get it. So s wants someone (in this case, the other agent h) to help s get salt. An action to reach salt and utterances "I want salt" and "Pass me salt" are invoked by dynamical control. The utterance "I want salt" is accounted for as a more reactive behaviour, as discussed later.

Initially the two agents s and h are jointly associated with the following constraints. For the sake of simplicity, we do not represent time explicitly, though of course we can in principle. The current treatment of time amounts to Agre and Chapman's deictic representation [1].

(1) true −happy(s).
(2) −haveSalt(s) +happy(s).
(3) −passSalt(X,Y) +haveSalt(Y).
(4) −saylWantSalt(X) −haveSalt(X) +happy(X).
(5) −sayPassMeSalt(X,Y) +passSalt(Y,X).

Atomic formula happy(X) means that agent X is happy. passSalt(X,Y) means that X passes salt to agent Y. saylWantSalt(X) means that X says "I want salt." Clause (4) describes the effect of an utterance "I want salt." That is, if X says the utterance, then X becomes happy after getting salt (happy(X)). These clauses are represented graphically as shown in Figure 2.

We tentatively assume for simplicity that different agents are associated with different dynamical systems. In Figure 2, enclosed by broken curves are the two parts of the constraint which the speaker and the hearer are attuned to. These parts are associated with different dynamical systems. Thus each atomic constraint in the overlapping area has two activation values.

First, we consider the speaker's dynamics and behaviour. The activation value of true in the top clause must be high. Due to exclusion energy, happy(s) is hence activated. Then, happy(s) in clause (2) and happy(X) in (4) are activated because of assimilation energy. In addition, haveSalt(s) in (2) and saylWantSalt(X,Y) and haveSalt(X) in (4) are activated because of disjunction energy. Similarly, the other constraints are also activated due to spreading activation. Since the activation

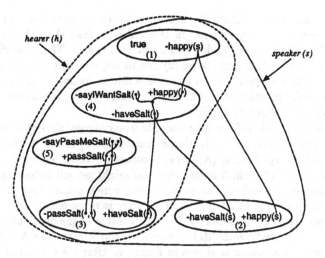

Fig. 2. Initial State of Constraint

value of haveSalt(X) in (4) is assumed to be low, a plan consisting of (2) and (3), which means that you can get salt if someone passes it to you, is preferred to a plan consisting of (4), which means that you might get salt by saying "I want salt." Then, action passSalt(s,s), unifiable with passSalt(X,Y) in (3), is executed. That is, it is introduced in the top clause as a literal −passSalt(s,s) as shown in Figure 3. In general, an action is executable only when its performer is the

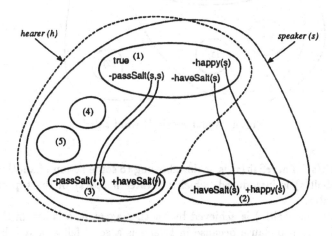

Fig. 3. Agent *s* got salt by herself.

agent in question[7]. passSalt(s,s) means that *s* passes salt to herself. If this action

---

[7] The repertoire of executable actions is built into the computational architecture.

causes achievement of haveSalt(s), then –haveSalt(s) is introduced in the top clause as shown in the same figure. In this case, haveSalt(s) in the top clause excites haveSalt(s) in clause (2) due to assimilation energy. Then, haveSalt(s) in clause (2) activates happy(s) in clause (2) due to disjunction energy. happy(s) in clause (2) activates happy(s) in the top clause due to assimilation energy. This increase the utility function U and there remains no more possible inferences or actions important enough to execute.

When passSalt(s,s) fails for some reason, however, +passSalt(s,s) is introduced in the top clause. Then, the activation of passSalt(X,Y) in (3) in the previous plan becomes low. sayIWantSalt(X) in (4) becomes activated higher than others. A plan consisting of (4), which means that you might get salt by saying "I want salt," is preferred to another plan including (5), which means that you can get salt by saying "Pass me salt," because the latter is more distant from the top clause and importance (measured by $\frac{\partial G}{\partial e}$) is larger due to the decay of backpropagation. So, action sayIWantSalt(s) is executed and literal –sayIWantSalt(s) is introduced in the top clause as shown in Figure 4. After the execution of action

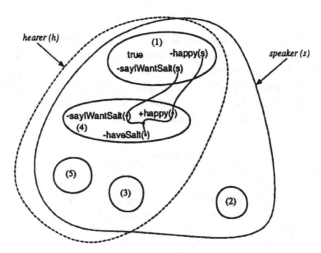

Fig. 4. Agent s said 'I want salt.'

sayIWantSalt(s), if haveSalt(s) is achieved owing to some action of another agent, then –haveSalt(s) is introduced, which activates happy(s) in the top clause as mentioned above.

In reality, haveSalt(s) is achieved by another agent h. h is assumed to have –happy(s) in the top clause because in brief s is a good fellow and h hopes for s's happiness. h's initial constraint network is shown in Figure 2. By hearing s's utterance "I want salt," –sayIWantSalt(s) is introduced in the top clause as shown in Figure 5. Due to assimilation energy, sayIWantSalt(X) in (4) is activated, and the activation of haveSalt(X) becomes high because of exclusion energy. haveSalt(X) and passSalt(X,Y) in (5) are highly activated because of assimilation

51

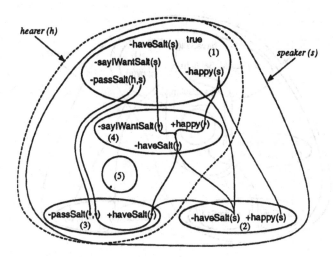

Fig. 5. Agent h heard 'I want salt.'

and exclusion energy, respectively. Then, since passSalt(X,Y) is executable when h is its performer, action passSalt(h,s), unifiable with passSalt(X,Y), is executed, and the literal −passSalt(h,s) is introduced in the top clause. After that, state haveSalt(s) is observed, then the literal −haveSalt(s) is introduced in the top clause as shown in the same figure.

Now let us return to agent s. If s is more deliberative, then s executes actions after more inferences have taken place. Since sayPassMeSalt(X,Y) in (5) causes the high activation of happy(s) in the top clause through clauses (3) and (2), action sayPassMeSalt(s,h) is invoked. Then, −sayPassMeSalt(s,h) is introduced in the top clause as shown in Figure 6, and the activation value of happy(s) increases. Furthermore, s's action sayPassMeSalt(s,h) being observed, the literal −sayPassMeSalt(s,h) is introduced as shown in this figure. Due to assimilation energy, the activation value of sayPassMeSalt(X,Y) in (5) becomes high, and passSalt(Y,X) is highly activated because of disjunction energy. $|\frac{\partial U}{\partial w}|$ is hence large and action passSalt(h,s), unifiable with passSalt(Y,X), is preferred. As a result, −passSalt(h,s) is introduced in the top clause.

As discussed above, the status of constraint varies according to the interaction between agents and their ever-changing environments, and reactive/deliberate behaviours emerge from the dynamical control of information flow. Of course, there are a tremendous number of clauses in an actual system of a highly intelligent agent along with the clauses introduced in the above example. We think that even though there are some clauses that are incompatible with the process in the above example, they are properly treated by dynamical interaction with the environment.

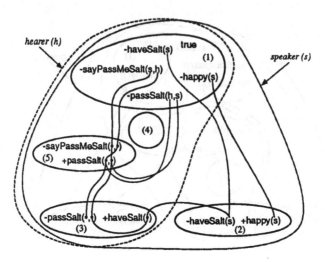

Fig. 6. Agent *s* said 'Pass me salt.'

5 Related Work

As already mentioned in Section 2, this work is closely related to the behaviour-based models such as subsumption architecture [4], situated automata [17], and activation/inhibition dynamics [14], in the sense that both concern embedding in dynamic environments without centralised control. Our work is unique in that declarative semantics and computational scheme are systematically related on the basis of dynamics to capture full first-order descriptive capability.

As Forrest states in the book 'Emergent Computation' [11], the essence of emergent computation or collective information processing is summarised as follows: (1) A collection of agents, each following *explicit instructions*; (2) Interactions among the agents (according to the instructions), which form implicit global patterns; (3) A natural interpretation of the epiphenomena as computations. Our approach also aims at realising emergent computation, but is distinguished in that it utilises the design orthogonal to information flow, and dispenses with not have even local explicit instructions. Our control of information flow emerges from dynamical interactions.

Eco-Problem Solver [10, 9] is one of the distributed problem solvers. Like other multi-agent systems, this system has domain-dependent procedures implemented as deterministic finite automata. Our system has also domain-dependent descriptions as constraints, but it is more flexible and robust because it is free from explicit control of computation.

Our approach has a lot in common with reactive planning [1] and anytime (time-dependent) planning [8, 3]. The difference between reactive and deliberative is caused by preference between symbolic computation and action execution. In our system, this preference is determined according to interactions with environments. If an action in a plan eliminates some precondition of another action in the plan, then the plan should be hard to generate. However, it does not mean that such plan *cannot* be generated. At first, another (maybe more reactive) plan will be constructed, but later, when the previous plan is known to be useless, preference on constructing more deliberative plans may become higher. Of course, even highly complicated (deliberative) plans can be constructed by using sufficient resources (e.g. time).

6 Concluding Remarks and Further Work

We presented *emergent planning* based on dynamical constraint programming. There are problems in conventional planning methodologies that explicitly control information flow in a fixed manner. Since information flow cannot be predicted in advance, domain/task specific control cannot capture planning in a dynamic environment. Our architecture is constraint-based, so it is free from domain/task specific stipulation of information flow, and defines declarative semantics of constraints by real-valued potential energy. The constraints are orthogonal to information flow and free from the concept of input/output. The dynamics derived from potential energy controls context-dependent information processing. This allows the system to process information based on the focus of attention and to balance the system's reactivity and deliberativity according to interaction between the system and the environment.

For further research, we are planning to formalise joint intentions and speech acts in multi-agent systems. Our formalisation will be simpler than that of Cohen and Levesque [6, 7] among others. The Cohen and Levesque's theory of agents can be considered not as individual constraints that exist in each agent but as social constraints over agents and the environment. There are two points in this discussion. First, each agent's internal information processing should be simpler because each agent's constraints are only a part of whole feedback loops that realise social constraints. Second, our fine-grained declarative semantics with non-monotonicity can simplify the theory of joint intentions and speech acts. There has been some research for formalising speech act theory by using non-monotonic theories [15, 2]. Since their theories are based on the conventional model theoretic semantics, however, they should always consider agents' information processing as being extremely deliberative. Our dynamical semantics can remedy this defect. We consider the joint intentions and speech acts as emergent phenomena of agents' communication including dynamical interaction with the environment.

References

1. P. E. Agre and D. Chapman, "Pengi: An Implementation of a Theory of Activity," In *Proceedings of the 6th National Conference on Artificial Intelligence (AAAI-87)*, 1987.

2. D. E. Appelt and K. Konolige, "A Nonmonotonic Logic for Reasoning about Speech Act and Belief Revision," In *Proceedings of the 2nd International Workshop on Non-Monotonic Reasoning*, Lecture Notes in Artificial Intelligence, vol.346, pp.164–175, Springer-Verlag, 1989.

3. M. Boddy and T. Dean, "Solving Time-Dependent Planning Problems," In *Proceedings of the 11th International Joint Conference on Artificial Intelligence (IJCAI-89)*, pp.979–984, 1989.

4. R. Brooks, "Intelligence without Representation," *Artificial Intelligence*, vol.47, pp.139–160, 1988.

5. R. Brooks, "Intelligence without Reason," In *Proceedings of the 12th International Joint Conference on Artificial Intelligence (IJCAI-91)*, pp.569–595, 1991.

6. P. R. Cohen and H. J. Levesque, "Persistence, Intention, and Commitment," In P. R. Cohen, J. Morgan, and M. E. Pollack (eds.), *Intentions in Communication*, pp.33–69, MIT Press, 1990.

7. P. R. Cohen and H. J. Levesque, "Confirmations and Joint Action," In *Proceedings of the 12th International Joint Conference on Artificial Intelligence (IJCAI-91)*, pp.951–957, 1991.

8. T. Dean and M. Boddy, "An Analysis of Time-Dependent Planning," In *Proceedings of the 7th National Conference on Artificial Intelligence (AAAI-88)*, pp.49–54, 1988.

9. A. Drogoul and C. Dubreuil, "Eco-Problem-Solving Model: Results of the N-Puzzle," In E. Werner and Y. Demazeau (eds), *Decentralized A.I. 3*, pp.283–295, Elsevier/North-Holland, 1992.

10. J. Ferber and E. Jacopin, "The Framework of Eco-Problem Solving," In Y. Demazeau and J.-P. Müller (eds), *Decentralized A.I. 2*, pp.181–193, Elsevier/North-Holland, 1991.

11. S. Forrest (ed.), *Emergent Computation*, MIT Press, 1991.

12. K. Hasida, "Dynamics of Symbol Systems: An Integrated Architecture of Cognition," In *Proceedings of the International Conference on Fifth Generation Computer Systems (FGCS-92)*, pp.1141–1148, 1992.

13. K. Hasida, K. Nagao, and T. Miyata, "Joint Utterance: Intrasentential Speaker/Hearer Switch as an Emergent Phenomenon," In *Proceedings of the 13th International Joint Conference on Artificial Intelligence (IJCAI-93)*, pp.1193–1199, 1993.

14. P. Maes, "Situated Agents Can Have Goals," In P. Maes (ed.), *Designing Autonomous Agents: Theory and Practice from Biology to Engineering and Back*, pp.49–70, MIT/Elsevier, 1991.

15. C. R. Perrault, "An Application of Default Logic to Speech Act Theory," In P. R. Cohen, J. Morgan, and M. E. Pollack (eds.), *Intentions in Communication*, pp.161–185, MIT Press, 1990.

16. F. J. Pineda, "Generalisation of Backpropagation to Recurrent and Higher Order Neural Networks," In D. Z. Anderson (ed.) *Neural Information Processing Systems*, pp.602–611, 1988.

17. S. Rosenschein, "Formal Theories of Knowledge in AI and Robotics," *Technical Report*, CSLI-87-84, Center for Study of Language and Information, Stanford University, 1987.

Emergence of sociality

Coalition formation among autonomous agents: Strategies and complexity (preliminary report) *

Onn Shehory   Sarit Kraus

Department of Mathematics and Computer Science
Bar Ilan University Ramat Gan, 52900 Israel
{shechory, sarit}@bimacs.cs.biu.ac.il
Tel: +972-3-5318863
Fax: +972-3-5353325

Abstract. Autonomous agents are designed to reach goals that were pre-defined by their operators. An important way to execute tasks and to maximise payoff is to share resources and to cooperate on task execution by creating coalitions of agents. Such coalitions will take place if, and only if, each member of a coalition gains more if he joins the coalition than he could gain before. There are several ways to create such coalitions and to divide the joint payoff among the members. Variance in these methods is due to different environments, different settings in a specific environment, and different approaches to a specific environment with specific settings. In this paper we focus on the cooperative (super-additive) environment, and suggest two different algorithms for coalition formation and payoff distribution in this environment. We also deal with the complexity of both computation and communication of each algorithm, and we try to give designers some basic tools for developing agents for this environment.

1   Introduction

Cooperation among autonomous agents may be mutually beneficial even if the agents are selfish and try to maximise their own expected payoffs [9, 20, 24]. Mutual benefit may arise from resource sharing and task redistribution. Coalition formation is an important method for cooperation in multi-agent environment. Agent membership in a coalition may increase the agent's ability to satisfy its goals and maximise its own personal payoff. There are two major questions concerning coalition formation: 1. which procedure should a group of autonomous agents use to coordinate their actions and cooperate; namely, how should they form a coalition? 2. among all possible coalitions, what coalition will form, and what reasons and processes will lead the agents to form that particular coalition?

---

* This material is based upon work supported in part by the NSF under Grant No. IRI-9123460.

Work in game theory such as [14, 17, 10, 12] describes which coalitions will form in N-person games under different settings and how the players will distribute the benefits of the cooperation among themselves. That is, they concentrate mainly on the second problem above. These results do not take into consideration the constraints of a multi-agent environment, such as communication costs and limited computation time.

In this paper we adjust the game theory concepts to autonomous agents, and present different coalition-formation procedures. The resulting coalitions of each of these procedures may be different even in the same setting. We developed general criteria for choosing between the coalition-formation procedures. Giving a specific setting and using these criteria will enable the agents to decide which coalition-formation procedure to prefer. We will concentrate on widely cooperative environments [4, 7] such as the postmen problem of [23]. The coalition-formation procedures that will be presented are either computation-oriented or negotiation-oriented. The appropriate procedure can be chosen according to the constraints of the environment and will lead to beneficial coalition formation.

We begin by describing the environment with which we will deal and give the basic definitions of coalitions in section 2.1. In section 3 we concentrate on super-additive environments. The negotiation protocol is described in section 3.1, together with its complexity analysis. The Shapley value protocol and its complexity are described in section 3.2. Finally, we give the criteria for choosing between the two protocols.

2 Environment description

This paper deals with autonomous agents, each of which has tasks it must fulfil and access to resources that it can use to fulfil these tasks. Resources may include: materials, energy, information, communications and others. Autonomous agents can act and reach goals by themselves, but may also join together to reach some or all of their goals. In such a case we say that the agents form a coalition.

In order to deal with different types of agents' environments, we give some general notations and definitions for concepts (i.e., resources, tasks, methods of reaching goals, success in reaching them, etc). Resources will be treated by numbers which will denote the quantities of each. Success in fulfilling tasks will be defined as the production or the payoff of an agent. The concept of payoff will be preferred specifically when reaching tasks is exchangeable with some kind of payment (e.g., money). The ways of reaching goals can be formalised by functions from the resources to the production or to the payoff. Each agent will have such a function of its own that tells what its way of using the resources to reach goals is.

An example of such an environment is the postmen domain [23]. While Zlotkin and Rosenschein consider only bi-agent environments (e.g., two postmen), we provide cooperation procedures for multi-agent environments (e.g., several agents, possibly more than two). In addition, Zlotkin and Rosenschein do not consider both the resource allocation problem and task distribution in the same setting e.g., we provide the postmen with procedures for division of their overall transportation capabilities and the benefits from fulfilling the letter distribution.

In order to make the creation of mutually beneficial coalitions possible, we make the following two assumptions:

Assumption 1 Communication
We assume that various communication methods exist, so that the agents can negotiate and make agreements [22]. We also assume that communications require time and effort on the part of the agent.

Assumption 2 Goods transferability & side-payments
We assume that resources and products can be transferred between agents in the environment, and that there is a monetary system that can be used for side-payments.

Multi-agents may reach agreements and form coalitions even if the second assumption is not valid (e.g., [10, 23, 1, 2]). However, the ability of goods transferability or side-payments may help the agents form more mutually-beneficial coalitions.

Our postmen example satisfies the assumptions above; they can communicate, and letters, transportation and money (their resources and production) can be transferred.

2.1 Definitions

Let N be a group of n autonomous agents, $N = \{A_1, A_2, \ldots, A_n\}$. Let L be a set of m resources, $L = \{l_1, l_2, \ldots, l_m\}$. The resources in the environment are limited. We consider only the resources that are distributed among the agents. This is formalised by the following definition:

Definition 1. Resource domain
We say that Q is the possible resource domain if Q is a set of the possible vectors of quantities of resources, $Q = \{\langle q^1, q^2, \ldots, q^m \rangle\}$, where q^j is a quantity of the resource l_j.
Each agent A_i has its own resources, therefore it has a vector $q_i \in Q$, $q_i = \langle q_i^1, q_i^2, \ldots, q_i^m \rangle$, which denotes the quantities of resources that A_i has. Due to the limitation of the quantities of the resources, we define $q_D \in Q$ to be the vector of the quantities of the resources of the domain. It is obvious that every element k of q_D, must satisfy $q_D^k = \sum_{j=1}^{n} q_j^k$, where q_j^k is agent A_j's quantity of resource k. The vector q_D denotes the total amounts of resources available to all agents in the environment, together.

We are interested in q_D particularly in cases of cooperation between agents, where q_D may sometimes be re-distributed among them. The agents use the resources they have to execute their missions. Depending on the quantities of resources an agent has and on its way of using them to fulfil tasks, each agent can reach all, part or none of its goals. For example, the postmen resources are the letters they have to distribute and their transportation abilities.

Production is a quantitative way of measuring the agent's success in fulfilling tasks. As such, we shall call production any kind of success in fulfilling tasks or reaching goals. Therefore, we give the following definition:

Definition 2. Production set
We say that P is a production set if P is a set of values which are the possible production values of a group of agents.

For example, if the agents are postmen, then the production set may be the set of amounts of letters that they can distribute, or their potential income.

Each agent in the environment has its own method of using the resources to reach goals. Therefore, we say that each agent A_i has a function f_P^i that formulates its way of using resources to reach goals; this function is defined as:

Definition 3. Production Function
We say that $f_P^i : Q \to P$ is a production function, if f_P^i is a function which gives a value within the production set that measures the production of agent A_i with an arbitrary resources vector.

It may be inconvenient to use the production function for agents' decision making. Each designer of automated agents has to provide its agent with a decision mechanism based on some given set of preferences [21]. Therefore, we suggest that each autonomous agent will be provided with a numerical payoff function that gives a transformation from the production set values and the resources to the reals. Each agent A_i has such a function, U^i, that exchanges its resources and production into monetary units. This monetary system can be used for evaluation of production and for side-payments.

Definition 4. Payoff Function
We say that $U^i : (P, Q) \to \mathcal{R}$ is a payoff function if U^i gives a measure, in monetary units, of the outcome that an agent has for some arbitrary resources and production.

A postman's payoff function is the difference between his transportation costs and the payment that he receives for distributing letters. The payments that postmen receive are transferable among them.

Individual self-motivated agents may be cooperative; they may cooperate by sharing resources, redistributing tasks and passing side-payments. Self-motivated agents will cooperate only if, as a result of this cooperation, they increase their payoff.

A coalition can be defined as a group of agents that decided to cooperate and also decided how the total benefit should be disbursed among its members. Formally, we define:

Definition 5. Coalition
Given a group of agents N, a resource domain Q, and a production set P, we

define a coalition as a quadrate $C = \langle N_C, Q_C, \bar{q}, U_C \rangle$. In this quadrate, $N_C \subseteq N$; $Q_C = \langle q^1, q^2, \ldots, q^m \rangle$, $Q_C \in Q$, is a coalitional resource vector, where $q^j = \sum_{A_i \in N_C} q_i^j$ is the quantity of resource l_j that the coalition has. $\bar{q}$ is the set of vectors of resource quantities after the redistribution of Q_C among the members of N_C ($\bar{q}$ satisfies $q^j = \sum_{A_i \in N_C} \bar{q}_i^j$). $U_C = \langle u^1, u^2, \ldots, u^{|C|} \rangle$, $u^i \in \mathcal{R}$, is the coalitional payoff vector, where u^i is the payoff of agent A_i after the redistribution of the payoffs.

We say that $\mathcal{C}$ is a set of possible coalitions if $\mathcal{C}$ is the group of all possible coalitions over N. In order to provide the agents with a method for coalition evaluation, we give each coalition a value (as can be found in game theory [12]), and a function for calculating this value.

Definition 6. Coalition Value & Coalitional Function
Let $C = \langle N_C, Q_C, \bar{q}, U_C \rangle$. We say that V is the value of C if the members of N_C can together reach a joint payoff V. That is, $V = \sum_{A_i \in N_C} U^i(\bar{q}_i)$, where U^i is the payoff function of agent A_i and $\bar{q}_i$ is its vector of resources after their redistribution in the coalition.

Let q_i be agent A_i's original vector of resource quantities; we assume that $\forall A_i$, $\bar{q}_i \in \bar{q}$, $u^i \geq U^i(q_i)$. That is, an agent will join a coalition only if the payoff it will receive in the coalition is greater than, or at least equal to, what it can obtain by staying outside the coalition. Hence, it is easy to conclude that $V \geq \sum_{A_i \in C} U^i(q_i)$. Moreover, we assume that the resources are redistributed within $\bar{q}$ in a way that maximises the value of the coalition. Therefore, the coalition value V of a specific group of agents N_C is unique. The complexity of computing the redistribution of the resources and calculating the coalitional value depends on the structure of the coalition's members' payoff functions. For example, for linear payoff functions the simplex method can be used.

Now, let us take the production function in definition 3 and expand its scope: We say that $f_P^C : \mathcal{C} \to P$ is a coalitional production function if f_P^C attaches a value $p \in P$ to all coalitions in $\mathcal{C}$. We say that $U^C : (P, Q) \to R$ is a coalitional payoff function if U^C transforms the coalitional production into coalitional payoff.

We now formally state more assumptions that consider a multi-agent environment in which the agents are self-motivated and try to maximise their own payoff function.

Assumption 3 Coalition joining (personal rationality)
We assume that each agent in the environment has personal rationality, i.e., it joins a coalition only if it can benefit at least as much within the coalition as it could benefit by itself. An agent benefits if it fulfils some or all of its tasks, or gets a payoff that compensates it for the loss of resources or non-fulfilment of some of its tasks[2].

---

[2] This assumption is usually called "personal rationality" in the game theory literature [8, 17, 12].

Assumption 4 Personal payoff maximisation
We assume that each agent in the environment tries to maximise its personal payoff; among all the possibilities that it has, an agent will choose the one that will give it the maximum expected payoff [3].

Our postmen example satisfies the two assumptions above: they are personally rational, and they try to maximise their personal payoff.

Now we shall define a new concept – the coalitional rationality. The environment is coalitionally rational if each coalition in it will add a new member only when the value of the coalition that will be formed by this addition is greater than (or at least equal to) the value of the original coalition. Not all environments are coalitionally rational. Situations come to mind whereby an agent is added to an existing coalition only because of the additional benefit for powerful agents and for the newly-added agent. Thus, the assumption of personal rationality is fulfilled, while the value of the new coalition is lower than the value of the original coalition. One can say that such a situation is not likely to happen, and can justify this statement by claiming that if some agents benefit more and the value of the new coalition is less than the value of the original one, then there must be at least one agent that will benefit less in the new coalition. That agent should never let such a coalition form, as the personal rationality assumption dictates. We reject this claim and say that such situations may happen if, after the change in the coalition, the agents that benefit less still prefer to stay in the new coalition, because they believe that this is the best option, given the new situation.

3 Super-additive environment

The concept of coalitional rationality can be broadened if we project it from the relation between a coalition and a single agent to the relation between two coalitions. By this expansion we define the super-additive environment:

Definition 7. Super-additive environment
A super-additive environment is a set of possible coalitions C that satisfies the following rule: for each pair of coalitions C_1, C_2 in the set C , $C_1 \cap C_2 = \phi$, if C_1, C_2 join together to form a new coalition, then the new coalition will have a new value $V_C^{new} \geq V_C^1 + V_C^2$, where V_C^1, V_C^2, V_C^{new} are the values of the coalitions.

Not all environments are super-additive, but if an environment is super-additive, then, under assumptions A1 - A4, after a sufficient time period a grand coalition will form. A grand coalition is a coalition that includes all of the agents. We come to the conclusion that a grand coalition will form because in any other situation there will be at least two coalitions that are not a grand coalition. If the environment is super-additive then the coalitional rationality and the personal

---

[3] Note that assumption 3 can be derived from assumption 4. We present here both assumptions only for clarification purposes.

payoff maximisation will make them join together and form a joint coalition, that, according to the super-additivity concept, will have a value $V^{new} \geq V^1 + V^2$ [18]. We can let the agents negotiate coalition formation, but all agents know that in a super-additive environment a grand coalition will form. Hence, the only problem that still remains is how the payoff should be distributed among its members. It seems that the best way to save time and effort for all agents will be to agree, without any argument, to form a grand coalition, and to solve the problem of payoff distribution either by negotiation or by calculation.

Definition 8. Common extra payoff
In a super-additive environment, if two coalitions C_i, C_j with corresponding coalitional values V_i, V_j can form a joint coalition and obtain together $V_{ij} = V_i + V_j + D_{ij}$, then D_{ij} is the common extra payoff (the indices i, j may be dropped).

We present two algorithms for solving the problem stated above: one that requires negotiations and another that requires only computations. We discuss the advantages of each later.

3.1 Negotiation in a super-additive environment

The initial coalitional state consists of n, single agent, coalitions. The coalitions then begin negotiating [11] and, step by step, form coalitions. Coalitions will continue negotiations via agents who are representatives of the coalitions of which they are members. These will form bigger coalitions, until a grand coalition forms. At the beginning of each step of this coalition formation process, each coalition will find what the common extra payoff from forming a new coalition with each of the other current coalitions is. Next, each coalition will sort its list of extra payoffs. The first coalition on each sorted list (i.e., the one that can bring maximum extra payoff) is "wanted" by the coalition that made this list. Two coalitions that want one another can start a bargaining process. In a super-additive environment, at least one such pair must exist. If there is more than one pair, then all will start a bargaining process. A bargaining process entails coalitions offering one another partition of their common D. A coalition that received such an offer may accept it, and then a new joint coalition can form. A coalition may also reject the offer, and either ask for a better one, or make its own offer. To establish order in the bargaining process, we define a strength relation between coalitions, and use this relation to determine which coalitions will be first to make offers.

Strong coalition We shall define a strong coalition by saying that coalition C_i is stronger than coalition C_j, if C_i has an extra payoff D_{ik} with some other coalition C_k, $k \neq j$, which is bigger than D_{ij}. It can be shown that two members of a pair of coalitions that are "wanted" by one another (such as that mentioned above), are both stronger than all other coalitions[4]; therefore we shall call them "powerful" [3]. We denote these two coalitions as C_p^1, C_p^2.

---

[4] Note that it can be that C_i is stronger than C_j and C_j is stronger than C_i.

The bargaining process is begun by the powerful coalitions. Within the pair of powerful coalitions, the coalition with the higher computational capabilities will be the first to calculate an offer and send it to its partner. We shall designate this first coalition as C_1^p and its associate as C_2^p. The first offers made to one another will be exactly half of D, their joint extra payoff[5]. Then, each powerful coalition will contact all coalitions that are weaker than it, inform them of the amount of the first offer it received and find among them the one coalition that is willing to give it the largest share of their common D. Any strong coalition, having received an offer (or offers) with δ higher than that presented by powerful coalition, will use the offer to challenge this powerful coalition. That is, the strong coalition will ask the powerful coalition for exactly as much of the common D as it could get from other coalitions.

At this point, the powerful coalition C_p^1 must contact its parallel coalition C_p^2, inform it about: 1. the amount of the highest offer it received, and 2. the coalition C_w^i that offered it. If C_p^2 was also challenged, and its highest offer was given by a coalition C_w^j, $i \neq j$, then C_p^1 and C_p^2 will reject one anothers' offers and each will join with its challenger. If both C_p^1 and C_p^2 are challenged by C_w^i, then C_p^1 has priority to join C_w^i and so C_p^2 should check if it has another challenging coalition. If so, then again, as above, both will reject one another. If C_p^2 was not challenged by the offers made, it should find the highest offer among these. If the difference between D_{12}^p and this highest offer is greater than C_p^1's highest offer, C_p^2 will accept the new offer and join C_p^1. Otherwise it will reject C_p^1 and C_p^1 will join C_w^i.

Weak coalition Each weak coalition, C_w, that was approached by another coalition, C_s, will calculate what it can offer by considering two issues: a. whether or not it is able to offer at least as much as C_s could get without cooperation with C_w; b. if the answer to (a) is positive, how much more it can offer. Consequently, C_w will follow this procedure: among all coalitions, excluding C_s, the coalition C_w will find those that are weaker or equal to it (the stronger ones will approach C_w). If there are none, then C_w will offer C_s as much as C_s claims it can get, plus ε (ε is a positive infinitesimally small number), up to a maximum of all of their common D_{ws}. This proposition is valid, unless D_{ws} is smaller than the offer that C_s already has – then C_w will have to offer D_{ws}, which will probably be rejected. If weaker coalitions exist, C_w will contact them, and each of these coalitions will have to go through the same procedure, recursively, to calculate their answer to the calling coalition. After C_w was answered by all of the contacted coalitions, it finds the one that offers it the maximum amount of payoff. This amount is the minimal part of D that C_w should ask for, from its common D with C_s, and the maximum is the amount that C_s claimed it could get, plus ε. To make it more clear, the following should be the typical answer of a requested coalition to its

---

[5] This partition is suggested only as a fair starting point of the negotiation. During the negotiation process the partition of D will change according to the relative strength of the coalitions.

caller: "You said that you can get x, and therefore I am willing to give you $x + \varepsilon$, but if you get a better offer, I can give you a maximum of $x + \delta$, and I will still be satisfied".[6]

Payoff disbursement When two coalitions C_1 and C_2 join together to form a new coalition, they get extra payoff D_{12}, which is divided between them into D_1, D_2, such that $D_1 + D_2 = D_{12}$. Now, we must present a method for distributing D_1 and D_2 among the members of C_1, C_2, respectively. Under the assumptions of coalitional rationality and super-additivity, we suggest that once a coalition forms, it will not split up. Therefore, in order to make these assumptions acceptable to the members of a coalition, we say that within an existing coalition, each member keeps its previous strength – the strength that was expressed by the ratio between the total D and the member's part of it when it joined the coalition. Therefore, any new added payoff should be divided among the coalition members according to their strengthes during the process of coalition formation. Of course, this does not prevent them from keeping for themselves any amount of payoff which they had before-hand.

Complexity of the negotiation algorithm The efficiency of this algorithm should be judged from two main perspectives: computations and communications. The first step of the negotiation process will be the transfer of all of the relevant information, i.e., payoff functions and resources vectors, between the agents (unless it is known in advance). This requires $(n-1)^2$ communication operations. The negotiation process will proceed from the initial state of n single agents to the final state of a grand coalition, and will take up to $n - 1$ steps, since at each step the number of coalitions will decrease by at least 1. At each step, each determines its extra common payoff with all other coalitions; this requires, $n - 1$ maximisation operations, which are rather complicated. Upon deriving this information, each coalition must sort it $(o(n \lg_2 n))$, and then contact the first coalition on the sorted list to find out if they are a powerful pair. Next, all strong coalitions contact all weaker coalitions, and all weaker coalitions, after having finished their calculations, re-establish contact with the corresponding strong coalitions in order to answer them. Altogether, there are $2(n-1)^2$ communication operations, and each such operation is connected with $o(1)$ calculations. Afterwards, the powerful coalitions contact one another once again, and search for the highest offer they received. At last, coalitions join together to form new ones. Since, in the end, all coalitions join together, there can be at least $\lceil \lg_2(n-1) \rceil$ and at most $n - 1$ such events. Any two coalitions that join together and form a new coalition, must perform the following actions:
a. choose a representative from among its members. This action will take $o(n)$ communication operations and the same order of computations.

---

[6] This assumes honesty or complete information on both the coalitional values and the messages that are transmitted. Otherwise, it will require long and exhausting negotiations to reach an agreement.

b. distribute the extra payoff among all members. With a given simple algorithm to do so, the complexity of this action is $o(n)$.

Now that we have determined the complexity of all the parts of the algorithm, we can construct the complexity of the general algorithm. There are at most $n-1$ negotiation steps; at each step there are $o(n^2)$ communication operations, and $o(n^2 \lg_2 n)$ computations. Therefore, the complexity of the general algorithm is, in the worst case, $o(n^3)$ communication operations, and $o(n^3 \lg_2 n)$ computations.

Example We shall use here an example to illustrate the negotiation algorithm. There are three agents in a super-additive environment. Each agent receives no payoff by itself, so that the coalitional values of a single-member coalition is zero. The values of the other coalitions are $V_{12} = 10$, $V_{13} = 7$, $V_{23} = 2$, $V_{123} = 15$. Agents A_1 and A_2 are the pair of powerful coalitions, since their common extra utility D_{12} is greater than their extra utility with A_3 (This is easy to conclude since in our specific example $\forall i, j$, $D_{ij} = V_{ij}$). Suppose that agent A_1 starts the negotiation by offering A_2 5, half of their common D. A_2 will offer A_1 the same amount. At this point A_1 will approach A_3, inform it about the current offer and ask for a better offer. Agent A_3 will offer A_1 $5 + \varepsilon$, and will announce that it can offer up to 7, if necessary. Meanwhile, Agent A_2 will also approach A_3. The offer of A_3 to A_2 will be 2. With the offer A_1 has received from A_3 it will approach A_2 and ask for more than 5, since it can receive more than 5 from A_3. Agent A_2 will compete with A_3's offer by offering A_1 $7 + \varepsilon$. At this point, none of the agents can make an offer that will change the payoffs and will be accepted. Therefore A_1 and A_2 will form a coalition; A_1 will receive a payoff of $7 + \varepsilon$ and A_2 will receive a payoff of $3 - \varepsilon$. Thus, one step of the negotiation is accomplished. If agent A_3 would not have indicated that he can offer up to 7, a possible scenario is that A_2 will offer A_1 $5 + \delta$ for some $\delta > \varepsilon$, and A_3 will respond by increasing its offer. This incremental process can proceed until eventually A_2 offers A_1 $7 + \varepsilon$. As we explained above, the fact that A_3 indicated that it can offer up to 7 shortened the bargaining process.

The next step of the algorithm will cause the formation of the grand coalition. Since there are only two coalitions at this stage, and their common extra payoff is 5, the only possible offer is to divide D by 2, so that each coalition will receive 2.5. This offer will be accepted, and the coalition of A_1 and A_2 will divide its new additional payoff according to the previous relative strengthes as was shown in the last step. The resulting payoff vector will be: $U_1 = 7 + 0.7 \times 2.5 + \varepsilon \approx 8.75$, $U_2 = 3 + 0.3 \times 2.5 - \varepsilon \approx 3.75$, $U_3 = 2.5$. Note that the final payoffs of the agents reflect their relative strength.

Discussion of the negotiation algorithm The outcome that a coalition gains as a result of the negotiation algorithm reflects its relative contribution to possible coalitions and its computational capabilities. Coalitions which contribute more relative to others will be preferred by the other coalitions and will receive a larger share of the common extra payoff.

As written above, when a pair of "powerful" coalitions is created, the coalition with the higher computational capabilities will be the first to calculate an offer and send it to its partner. The situation of being the first to make an offer in the negotiation process may give coalitions an advantage when joining with computationally weaker coalitions. This means that the algorithm may be advantageous to agents that have better computational capabilities.

There may be occasions when a coalition A obtains its highest common extra payoff with more than one other coalition. If A's highest common extra payoff with two different coalitions B_1 and B_2 is identical, then A should choose one of them to be its partner in a "powerful" pair. Being chosen by A may give the chosen coalition an advantage over the other. The ability to choose from among B_1 and B_2 the one that will be included in a pair provides A with some advantage in the negotiation. For example, A can demand to be designated as C_p^1, i.e., to be the first to make an offer when negotiating the formation of a joint coalition. One possible criteria that A may use for choosing among B_1, B_2 is their computational power. The algorithm requires that after a coalition has chosen its partner in a pair, it cannot depart. This requirement leads to stability of the negotiation process.

Being chosen as a "powerful" coalition according to the algorithm does not necessarily give an advantage to the chosen coalition. This is because a powerful coalition C_p has to approach a weaker coalition C_w. The proposal that C_w returns to C_p may be the minimum necessary for cooperation with C_p, although C_p can decide to avoid cooperation with C_w and cooperate with another weak coalition or with its associate, the other powerful coalition. The main factor that plays a role in the coalition's eventual outcome is its relative contribution to possible coalitions. Whether being "powerful" is advantageous for the powerful coalition depends on the details of the specific situation, and cannot be predicted without complex computation.

Although powerful coalitions do not necessarily have an advantage over other coalitions, in the negotiation algorithm we do prefer that these coalitions, which have a larger common extra payoff, will begin the negotiation process. This is because coalitions with larger common extra payoffs will have more flexibility when negotiating coalition formation. Therefore, it is more likely that such coalitions will form joint coalitions faster and with less computations than other coalitions. Thus, the overall costs will decrease, an advantage that the designers of agents should seek.

3.2 Shapley value

A well-known solution to the problem stated above was suggested by Shapley[7] [18]. Given a group of agents in a super-additive environment, and assuming that a grand coalition had formed, Shapley suggests a function φ that attaches to each agent A_i in the grand coalition a unique value v_i, which is its share of

---

[7] For more discussion on Shapley see [13, 6, 5].

the joint payoff. φ is a general payoff function, which is defined for all of the agents. Shapley suggested that φ should satisfy the following axioms:

Axiom 3.1 Symmetry
For every pair of agents $A_i, A_j \in N$, for all the coalitions such that $A_i, A_j \notin N_C$, if $V(N_C \cup \{A_i\}) = V(N_C \cup \{A_j\})$ then $v_i = v_j$, and we say that A_i, A_j are exchangeable.

Axiom 3.2 Null agent
For every agent $A_i \in N$, for all the coalitions such that $A_i \notin N_C$, if $V(N_C \cup \{A_i\}) = V(N_C)$ then $v_i = 0$, and we say that A_i is null.

Axiom 3.3 Efficiency
For all the agents $A_i \in N$, $V(N) = \sum_i v_i$.

Axiom 3.4 Additivity
For every pair of coalitions $C_i, C_j \in C$, such that $N_C^i = N_C^j = N_C$, if Q_C^i, Q_C^j and $\bar{q}^i, \bar{q}^j$ give these coalitions the values V^i, V^j, then the value of the coalition $C_{ij} = \langle N_C, Q^i + Q^j, \bar{q}^{ij}, U_C^{ij} \rangle$ is $V^{ij} = V^i + V^j$.

Shapley proves [18] that the four axioms above are enough for determining φ uniquely, and also gives an explicit formula for calculating φ.

We shall denote by $\mathcal{R}$ a permutation of the members of N. It is obvious that there are $n!$ different permutations. We shall denote by $C_i^{\mathcal{R}}$ the group of agents that were included in $\mathcal{R}$ before agent A_i. Thus, the expression $V(C_i^{\mathcal{R}} \cup \{A_i\}) - V(C_i^{\mathcal{R}})$ is the marginal payoff that agent A_i brings to the coalition $C_i^{\mathcal{R}}$. If we sum agent A_i's marginal payoff over all possible $\mathcal{R}$ and calculate the mean by division by $n!$, then we obtain Shapley's formula:

$$\varphi(A_i) = \frac{1}{n!} \sum_{\mathcal{R}} [V(C_i^{\mathcal{R}} \cup \{A_i\}) - V(C_i^{\mathcal{R}})]$$

We suggest a method for using Shapley's formula:
Assuming honesty, we can randomly choose an agent A_r that will be responsible for calculating the Shapley values (A_r may be paid for its efforts). A_r may also be elected by some voting procedure [15, 16]. To be able to calculate the Shapley values, A_r must first contact all other agents and ask for all of the relevant information. Therefore it contacts $n - 1$ agents, which respond by transmitting their payoff functions and their resource vectors. After A_r has received all of the information, it computes the Shapley values; a process which requires the computation of all 2^n possible coalitions of the agents. For each possible coalition, A_r must calculate the corresponding payoff. This calculation is a maximisation of a function of several variables, which is rather complicated. When the computation is complete, A_r once again contacts all of the agents and informs them of the results. That is, A_r directs all agents how to divide the resources, tasks and extra payoff among them.

Complexity of the Shapley algorithm The Shapley algorithm requires $o(n)$ communication operations to be performed in the first stage. The number of computations for the calculation of Shapley values is dictated by 2^n, and hence the computational complexity is $o(2^n)$. After calculating these values, A_r must contact all of the agents; this requires $o(n)$ communication operations and the same order of computations. Altogether, the complexity of the Shapley algorithm is $o(n)$ communication operations, and $o(2^n)$ computations. A new formula for the Shapley value that offers a reduction in computation time was presented by [5], but it remains of order $o(2^n)$.

Example Recall of the last example of the three agents in a super-additive environment. Using Shapley's formula to calculate the payoffs of the agents in the grand coalition yields the following results: $\varphi(A_1) = 7.167$, $\varphi(A_2) = 4.667$, $\varphi(A_3) = 3.167$. These results are different from the results of the negotiation algorithm example, but the difference is small with respect to the payoffs. We can indicate that in this specific case, the negotiation algorithm strengthens the stronger and weakens the weaker, while the Shapley formula tends to impose "fairness". However, in other examples the negotiation algorithm may strengthen the weaker and weaken the stronger.

3.3 Discussion

Previously we presented two algorithms for payoff distribution among the members of a grand coalition in a super-additive environment. If communication is expensive compared to computation and all designers agree that a "fair" payoff distribution function should satisfy axioms 1 – 4, then it is worthwhile for all designers to agree, in advance, to use the Shapley algorithm. We derive the first condition (communication is expensive compared to computation) by comparing the complexities of the two algorithms. It is obvious that the Shapley algorithm is better only when computation is cheap in comparison with communication.

From the above, it follows that designers have some information about when to choose the Shapley algorithm. We must now compare it to the negotiation algorithm. Both the Shapley algorithm and the negotiation algorithm start with transmission of information; at this stage, the negotiation algorithm requires $o(n^2)$ communication operations, whereas the Shapley algorithm requires only $o(n)$ communication operations. Next, both algorithms require calculations of common payoffs of coalitions; each of these calculations is a maximisation over many variables, which is quite a complex calculation. In the Shapley algorithm, the common payoff calculation must be done for all 2^n possible coalitions, and all of these calculations are performed by one single agent in sequence. In the negotiation algorithm, common payoff calculations should be done only for coalitions that form during the negotiation process; that is, only $o(n^3)$ such calculations. Moreover, these calculations are distributed among the agents. This distribution is "natural" in the sense that it is an outcome of the algorithm characteristics, i.e., each agent performs only those calculations that are required for its own actions during the process. In contrast, the Shapley algorithm does not enable distribution of the calculations, mainly because of the ultimate need for agreement upon the redistribution of resources and payoff; if calculations are distributed, agreement requires negotiation.

An important advantage of the negotiation process is the ability to suspend it before it ends (i.e., anytime algorithm). If this algorithm stops before completion, then there is already a coalitional configuration which is for all agents at least equal to, and probably better than, their initial condition. It is obvious that such termination of the Shapley algorithm will bring the agents to the starting point and no coalition will form.

The negotiation algorithm can be performed without real negotiations, if we let one agent simulate the negotiation process and make all of the calculations. In this case the number of calculations will not change, but they will no longer be distributed. The number of communication operations will decrease to $o(n)$, exactly as in the Shapley algorithm.

The Shapley solution takes all possible negotiation scenarios into consideration while the negotiation algorithm follows only one scenario. Hence, we expect that the resulting payoff distributions of the two algorithms will rarely be equal (we have explicitly shown it for $n = 3$). It is hard to predict which algorithm will grant more to a given agent. Therefore, we cannot give an agent a method that will enable it, for any given setting, to determine which of the two methods it should prefer. Hence, we suggest to the designers of agents to take into consideration the costs of both communication and computation, and the significance of computation distribution.

The procedures presented above were developed for super-additive environments where forming bigger coalitions is always beneficial. However, in environments where there are unresolvable conflicts between the agents, forming bigger coalitions may not be beneficial or may even be harmful to some agents. Such environments are non-super-additive environments.

The Shapley value algorithm is designed particularly for super-additive environments and cannot be adjusted to non-super-additive environments. However, the negotiation algorithm can be adjusted, by making small changes, to environments which are not super-additive environments. Note that in several such situations the grand coalition will not form. This is beyond the scope of this paper. Non-super-additive environments are discussed in [19].

4 Conclusion

This paper discusses multi-agent environments where agents are designed to reach goals that were pre-defined by their operators. An important way to execute tasks and to maximise payoff is to share resources and cooperate on task execution by creating coalitions of agents. The paper discusses the advantages of two methods of coalition-formation and payoff distribution in super-additive environments and suggests occasions when each is most suitable. Both algorithms ultimately lead to grand coalition formation and payoff disbursements among coalition members. The algorithms are: 1. the Shapley value algorithm, which employs one representative, is computation-oriented and is best used in instances where communication is expensive. If calculations are halted in the middle, the process will not lead to any coalition formation, and; 2. the negotiation algorithm, which leads to distribution of both calculations and communications, and is primarily communication-oriented. If halted in the middle this algorithm still provides the agents with a set of formed coalitions.

References

1. R. J. Aumann. The core of a cooperative game without side-payments. *Transactions of the American Mathematical Society*, 98:539–552, 1961.

2. R. J. Aumann and B. Peleg. Von Neumann-Morgenstern solutions to cooperative games without side-payments. *Bulletin of the American Mathematical Society*, 66:173–179, 1960.

3. C. Castelfranchi. Social power. In Y. Demazeau and J. P. Muller, editors, *Decentralized A.I.*, pages 49–62. Elsevier Science Publishers, 1990.

4. R. Conte, M. Miceli, and C. Castelfranchi. Limits and levels of cooperation: Disentangling various types of prosocial interaction. In Y. Demazeau and J. P. Muller, editors, *Decentralized A.I. - 2*, pages 147–157. Elsevier Science Publishers, 1991.

5. G. Gambarelli. A new approach for evaluating the Shapley value. *Optimization*, 21(3):445–452, 1990.

6. S. Guiasu and M. Malitza. *Coalition and Connection in Games*. Pergamon Press, 1980.

7. J. C. Harsanyi. A simplified bargaining model for n-person cooperative game. *International Economic Review*, 4:194–220, 1963.

8. J. C. Harsanyi. *Rational Behavior and Bargaining Equilibrium in Games and Social Situations*. Cambridge University Press, 1977.

9. S. Kraus. Agents contracting tasks in non-collaborative environments. In *Proc. of AAAI93*, pages 243–248, France, 1993.

10. S. Kraus and J. Wilkenfeld. Negotiations over time in a multi agent environment: Preliminary report. In *Proc. of IJCAI-91*, pages 56–61, Australia, 1991.

11. T. Kreifelts and F. Von Martial. A negotiation framework for autonomous agents. In *Proc. of the Second European Workshop on Modeling Autonomous Agents in a Multi Agent World*, pages 169–182, France, 1990.

12. R. D. Luce and H. Raiffa. *Games and Decisions*. John Wiley and Sons, Inc, 1957.

13. M. Maschler and G. Owen. The consistent Shapley value for games without side payments. In R. Selten, editor, *Rational Interaction*, pages 5–12. Springer-Verlag, 1992.

14. J. Von Neumann and O. Morgenstern. *Theory of Games and Economic Behavior*. Princeton University Press, Princeton, N.J.

15. B. Peleg. Consistent voting systems. *Econometrica*, 46:153–161, 1978.

16. B. Peleg. *Game Theoretic Analysis of Voting in Committees*. Cambridge University Press, Cambridge, 1984.

17. A. Rapoport. *N-Person Game Theory*. University of Michigan, 1970.

18. L. S. Shapley. A value for n-person game. In H. W. Kuhn and A. W. Tucker, editors, *Contributions to the Theory of Games*. Princeton University Press, 1953.

19. O. Shehory and S. Kraus. Feasible formation of stable coalitions in general environments. Technical Report, Institute for Advanced Computer Studies, University of Maryland, 1994.
20. K. P. Sycara. Persuasive argumentation in negotiation. *Theory and Decision*, 28:203–242, 1990.
21. M. Wellman and J. Doyle. Modular utility representation for decision-theoretic planning. In *Proc. of AI planning Systems*, pages 236—242, Maryland, 1992.
22. Eric Werner. Toward a theory of communication and cooperation for multiagent planning. In *Proceedings of the Second Conference on Theoretical Aspects of Reasoning about Knowledge*, pages 129–143, Pacific Grove, California, March 1988.
23. G. Zlotkin and J. Rosenschein. Negotiation and task sharing among autonomous agents in cooperative domain. In *Proc. of the nth International Joint Conference on Artificial Intelligence*, pages 912–917, Detroit, MI, 1989.
24. G. Zlotkin and J. Rosenschein. Cooperation and conflict resolution via negotiation among autonomous agents in noncooperative domains. *IEEE Transactions on Systems, Man, and Cybernetics, Special Issue on Distributed Artificial Intelligence*, 21(6):1317–1324, December 1991.

Coalition Formation Among Autonomous Agents

Steven P. Ketchpel[1]

Stanford University, Computer Science Department, Stanford, CA 94305, USA

Abstract. Coalitions of agents can work more effectively than individual agents in many multi-agent settings. Determining which coalitions should form (i.e., what agents should work together) is a difficult problem that is typically solved by some kind of centralised planner. As the number of agents grows, however, reliance on a central authority becomes increasingly impractical. This paper formalises the coalition formation problem in decision theoretic and game theoretic terms and presents a fully distributed algorithm that can efficiently determine coalitions that will be approximately "stable." Stable coalitions are resistant to attempts of outsiders to break the coalition, because remaining in the coalition maximises the expected reward for each agent in the coalition. The algorithm is a variant of the "stable marriage matching with unacceptable partners" [6]. The Shapley value ([11], [12]) is suggested as a fair method to divide the coalition's utility among the members.

1 Introduction and Related Work

1.1 Introduction

The proliferation of computer systems has led to a new conception of how computers might work together. The so-called "open systems" structure looks forward to an environment housing a large number of systems of different designs that can work together. Given the broad range of tasks that may be assigned to these systems, flexible schemes of communication and co-ordination are required. The dynamic formation of coalitions is one such scheme.

Work in DAI has been split into two categories: multi-agent systems and distributed problem solving [9]. The algorithm presented below falls into the multi-agent systems area. It is concerned with the interaction of a large number of self-interested agents. Each agent is "rational" in the economic sense of being a utility maximiser. An independently motivated agent may not be willing to settle for a plan generated by a centralised planner. In many cases the community-optimal plan calls for one or more agents to sacrifice individual utility for the sake of the community. A rational agent would attempt to locate other plans that would maximise its own utility rather than executing the prescribed plan from the centralised planner. In addition, a central planning node is a potential communication bottleneck or system crippling fault point. Therefore, a decentralised solution is preferable, especially in a situation where the agents may be mistrustful of one another.

Agents in a multi-agent planning system may benefit by dynamically forming and dissolving work groups. This benefit may result from one of several reasons: 1) the agents may have partially or fully overlapping goals; 2) the agents may have different abilities, so that by exchanging tasks or resources, the assignment among the larger group may be more efficient; or 3) the agents may place different relative values on present and future gains.

---

[1] Author's e-mail address: ketchpel@cs.stanford.edu. The work was performed while the author was at Harvard University.

The present work addresses this problem of forming work groups or coalitions. More specifically, if each agent has its own goals, how do agents locate other agents with whom they can beneficially collaborate, and how can they fairly share the joint gain that accrues to the coalition? Formally, we want to partition the set of agents into subsets, each of which is a coalition. . Each coalition obtains a certain utility, which is shared among the members of the coalition.

1.2 The Desiderata of the Solution:

An ideal solution to this problem would have the following characteristics (adapted from [13] and [9]):
- Stability: Each agent is happy with its position, in that there is no way that agents outside a coalition could break the coalition apart by enticing coalition members away with better offers. The excluded agents must lack either the personal incentive to break it apart (i.e., forming a different coalition with some members of the current one would not be to the excluded agents' benefit,) or the means to lure away coalition members (i.e., an agent currently in the coalition would not expect a higher reward if it left). Furthermore, each agent that is included in a coalition prefers being in that coalition to working alone.
- Efficiency: The coalition formation process is efficient in terms of both computational cost and number of communications among the agents. If the added gain of collaborating with the other agents does not outweigh the cost of establishing the coalition, agents will continue to work independently.
- Decentralisation: Each agent should take part directly in the calculation and communication. This property is desirable to prevent reliance on a centralised manager, which would increase both the fragility of the system and the potential for bottlenecks.
- Symmetry: No special computational demands should be made on any one agent. All agents play a comparable role, and new agents can be introduced to the community of agents without disrupting the system.

1.3 Related Research

Ephrati and Rosenschein [2] use the Clarke Tax from economic theory to determine which agents will have their goals met, and how much they will be charged for that privilege. Although their agents have individual goals, there is a centralised decision procedure that arbitrates among the goals after obtaining information from all of the agents. The present work builds on previous work [7] which proposes a fully distributed mechanism for exchanging labour among the agents to exploit the different abilities of the agents. The algorithm presented there is based on the contract net [1] but ignores questions of stability. Other market-oriented approaches (e.g., [14]), match producers and consumers of services, but allow only well specified transfers of goods among them, falling short of the notion of forming a collaborative group. Grosz and her colleagues ([4], [5], [10]) formalise the notion of a collaborative plan, but assume the presence of agents that are willing to collaborate and do not address the choice of partners.

The use of decision theory techniques to analyse AI problems dates back to [3]. The Shapley value ([11], [12]) is a tool from game theory which is used here to determine a fair division of the utility among the agents in a coalition.

Gusfield and Irving [6] present the stable marriage algorithm which guarantees stable matchings among a set of men and women with ranked preference lists. They also include a variant of that algorithm called the stable roommate algorithm which is designed for a single class of agents that both makes and receives proposals. The algorithm is quadratic, and is guaranteed to find a stable matching if one exists, but it requires a centralised planner with complete knowledge of all of the agents' preference lists. In order for this approach to work, all of the agents would need to send their preference lists to one agent, thereby introducing a potential bottleneck and failure point. Additionally, agents may have privacy concerns that prevent the disclosure of their entire preference list. The central agent also makes changes in the preference lists, requiring that there be absolute trust in the central planner, because it is making decisions for the agents without consulting them. Therefore, the Gusfield and Irving proposal does not meet the design criteria.

2. The Algorithm

2.1 Overview

This section describes the coalition formation process, which consists of four phases: Communication, Calculation, Offers, and Unification. Figure 1 briefly describes each phase, and Sects. 2.3 through 2.6 provide further details. Taken together, these four phases comprise one round of coalition formation. Each round is limited to joining pairs of entities. Therefore, multiple rounds are required to form larger coalitions. Any coalition that forms in one round is treated as a single entity in subsequent rounds and plays the same role that agents play in the first round. Since some entities may emerge from a given round unpaired, there is the possibility that coalitions may form which have sizes that are not a power of 2.

1. Communication Phase: Each agent gathers the information it needs to determine the compatibility of potential coalition partners.

2. Calculation Phase: Each agent ranks the potential coalition partners, and determines a fair division of the joint utility.

3. Offers Phase: Agents proceed down the preference lists they constructed in Phase 2, extending offers, and accepting or rejecting offers that they receive.

4. Unification Phase: Pairs of agents that accepted each other's offers unify into a single entity for future rounds of formation.

Fig. 1. Phases in coalition formation process

The major contributions of this paper are in the second and third phases, applying the Shapley value for the division of the coalition's utility, and providing a fully distributed algorithm for efficiently creating coalitions of agents.

2.2 Assumptions

Several simplifying assumptions will be made in the definition of the problem. They are:
• Each agent can communicate with every other agent.
• Agents reveal only true information and are obligated to keep the offers they extend to coalition partners.
• Agents are synchronised, with messages being exchanged in discrete "rounds."
• The communication channels are perfect, so that all messages are transmitted faithfully and arrive at their destinations in the time step following the one in which they were sent.
• Utility units are comparable between agents, and further, they are transferable among agents.
In addition to the above simplifying assumptions, the following restrictions are made to increase the realism of the scenario.

- Communication occurs between two agents at a time. There is no mechanism for broadcasting information to all agents.
- Communication bandwidth is limited. Agents may only send and receive one message at a time. Additional messages are stored in a buffer.

2.3 Communication Phase

Communication among the agents allows them to locate other agents that may have compatible or overlapping goals, or agents that may have complementary skills. During the communication phase, each agent communicates with every other agent to determine how much synergy would be created by the formation of that pair of agents. Several values are needed for the agents to determine the joint utility. Using game theory notation, these values are $v(A)$, $v(B)$, and $v(AB)$: the value to agent A alone, the value to B alone, and the maximum value that is guaranteed for the two entities if they work together, even if all of the agents outside the coalition conspire against them. The content of the communication from agent S to agent R, then, consists of two pieces: 1) the precalculated $v(S)$; 2) the data that R needs to calculate $v(RS)$.

2.4 Calculation Phase
Once each agent has the requisite information from all of the community members, the calculation phase begins. The field of game theory offers several suggestions for criteria that may be used to divide the group utility. Simply splitting the utility equally among the agents in the coalition does not recognise the possibility that the agents may be making unequal contributions.

The Shapley value takes into account the different contributions of the agents and splits the reward accordingly. Selecting the Shapley value also ensures the sum of the utility shares of the coalition members is exactly the combined utility of the coalition. As a result, no further negotiation about the excess is required; also, it is impossible for the coalition to fail to obtain enough utility to pay all of the member shares, assuming that the agents agree on the values of the game, and that their valuations are the true ones. For the effects of weakening these assumptions, see [8].

The Shapley value is calculated by looking at each of the different dynamics that could lead to the coalition under consideration. The assumption is that agents form a coalition either by being the "founding" member, or else by joining, one at a time, a coalition that already exists. So the set of formation dynamics is simply the permutations of the agents in the coalition. Each agent adds value to a given formation process of the coalition based on the marginal utility gain created by that agent. For example, if agent A is joining agents B, C, and D, and $v(ABCD) = 50$ and $v(BCD) = 35$, then A's marginal contribution under this formation ordering is $v(ABCD) - v(BCD) = 15$. If agent A joins a coalition started by B, and then they are joined by agents C and D, A's marginal contribution is $v(AB) - v(B)$. These are only 2 of the 24 different permutations that might lead to the final coalition ABCD. By averaging A's marginal contribution across all the different formation possibilities, A's Shapley value is obtained. The underlying assumption is that all of the different formation processes are equally likely, and therefore, the marginal contributions for each formation are weighted equally. This calculation ensures that the sum of the Shapley values for all of the members of the coalition will be exactly the coalition's combined utility.

In general, the Shapley value requires looking at all of the permutations, an exponential operation. By limiting the formation to pairs of agents only, the expense of computing the Shapley value is a small constant. Computing the Shapley value for the formation of a two member coalition by agents X and Y requires only $v(X)$, $v(Y)$, and $v(XY)$. In this degenerate case of two entities forming a coalition, the Shapley value of X is:

$$\frac{1}{2}v(X) + \frac{1}{2}(v(XY) - v(Y))$$ (1)

symmetrically, the Shapley value of Y is:

$$\frac{1}{2}v(Y) + \frac{1}{2}(v(XY) - v(X))$$ (2)

Note that the sum of these two values is $v(XY)$.

Each agent creates a preference ordering among the offers, in order of decreasing utility that it expects to receive, including all of the agents that made an offer yielding a positive utility. There may be cases in which a coalition obtains a combined utility that is less than the individual utility obtainable by one or both of the coalition members. In this case, the agents prefer to remain unattached rather than forming a coalition, so these agents are not listed in the preference lists. They correspond to unacceptable partners in the stable marriage problem.

2.5 Offers Phase

The "stable marriage" algorithm [6] serves as a departure point for the coalition formation algorithm. The stable marriage algorithm with unacceptable partners is a generalisation of the stable marriage algorithm where the matching need not be total. An equal number of men and women participate. Each man has a ranked list of preferences among the women, and similarly, each woman has a ranked list of preferences among the men. Certain men may be "unacceptable" to a given woman, so her preference list may include only a subset of the men, and she would prefer being unmatched over a pairing with any "unacceptable" man. Men may also omit women from their preference lists because they are "unacceptable." A matching is a list of pairs of people that are associated. Each person may have at most one partner. A matching is stable if there is no blocking pair, where a blocking pair is defined as follows:

> There is a person A such that A prefers some person B to its current partner, and person B prefers A to his or her current state in which he or she is either matched to a lower preference partner or unpaired,
> OR
> Some person prefers being alone to his or her current partner.

The stable marriage with unacceptable partners algorithm proposed in [6] is $O(n^2)$, and guarantees a stable matching.

However, when the two classes ("men" and "women") are conflated to a single class of "agents," certain modifications are required. The efficiency of the modified algorithm is also $O(n^2)$, and the modified algorithm is also distributed over symmetric agents. The modified algorithm does not, however, ensure that the matching will be stable, even when stable matchings exist. It will be argued in Sect. 3.2 that the formation process effectively minimises the impact of this instability by discouraging agents from attempting to break up coalitions.

In the algorithm presented here, each agent lists in its preference list all of the other agents with which it could beneficially collaborate. The agents then extend offers to each other according to their preference lists, accepting offers that improve their positions, declining others, until the situation stabilises.

Figure 2 gives the pseudo-code implementation of the modified stable marriage algorithm from a centralised point of view. The agents actually execute a somewhat different algorithm, which is shown in Fig. 3. The net result duplicates the behaviour of the centralised algorithm, but with a distributed implementation. It is as if the centralised algorithm is called with the agent-list consisting of all the entities in the community.

Both the centralised and distributed algorithms make use of the "Next-Agent-to-Ask" and "Current-Partner" properties of an agent, which correspond to how far down in its preference list an agent is and what its current status is, respectively. In the distributed case, an agent only has access to the values of its own properties. At the termination of either algorithm, the "Current-Partner" property will determine which pairs of entities will form coalitions in that round.

If an agent P accepts the offer of another agent A, P may continue to make offers. Even though P prefers A to its current position, there may be an entity higher in P's preference list that would also like to be paired with P. Therefore, P will ask all of its choices ranked more highly than A before the offers phase terminates. For example, agent A might propose to agent P, and agent P accept but try to better its position. P might extend an offer to a third party S, and leave A if S accepts. If S subsequently leaves P, P will continue down its preference list, potentially asking A to accept on the same terms that A had previously offered, and P spurned.

```
STABLE (agent-list)
BEGIN

    IF agent-list is empty, RETURN.

    A := agent from the agent-list.
    REPEAT
      IF at end of preference-list(A) /*A prefers going solo*/
          STABLE(agent-list - A).
          RETURN.
      END-IF.

    A offers to join with the next agent on preference-list(A) call it P.

      IF P prefers A to its current status /*whether solo or another partner*/
          P accepts A's offer to join.
          IF Current-Partner(A) not nil
            agent-list := agent-list ∪ Current-Partner(A).
            Current-Partner(Current-Partner(A)) := nil.
          END-IF.
            Current-Partner(Current-Partner (P)) := nil.
            agent-list = agent-list ∪ Current-Partner(P).
            Current-Partner(P) := A.
            Current-Partner(A) := P.
            STABLE (agent-list - A).
            RETURN.
      ELSE
          P Rejects A's offer.
          Advance the position in the preference-list(A) for the next agent to ask.
      END-IF.
    UNTIL false.
END.
```

Fig. 2. The centralised version of the modified stable marriage algorithm

The decentralised algorithm requires synchronisation among the agents, which is captured in the variable MESSAGE-SYNC of the algorithm in Fig. 3. MESSAGE-SYNC goes through the values: EXTEND-OFFERS, RECEIVE-OFFER/SEND-RESPONSE, RECEIVE-RESPONSE, SEND-DUMP-MESSAGE-1, RECEIVE-DUMP-MESSAGE-1, SEND-DUMP-MESSAGE-2, RECEIVE-DUMP-MESSAGE-2. The round may only end if MESSAGE-SYNC has the EXTEND-OFFERS value.

```
BEGIN.
    SWITCH MESSAGE-SYNC
    CASE: "EXTEND OFFERS"
        IF (Current-Partner is preferred to Next-Agent-to-Ask)
            listen for end of phase
        ELSE
            IF NOT(offer-outstanding)
                extend offer to Next-Agent-to-Ask
                offer-outstanding := TRUE
            ELSE
                SEND "I'm not done yet!"
            END-IF.
        END-IF.
    CASE: "RECEIVE OFFER & SEND RESPONSE"
        get earliest offer from message buffer, call sender S
        IF S is preferred to Current-Partner
            DUMP1 := Current-Partner
            Current-Partner:= S
            SEND "I Accept" to S
        ELSE
            SEND "I Decline" to S
        END-IF.
    CASE: "RECEIVE RESPONSE"
        IF response in message buffer
            get response from message buffer, call Sender S
            IF response = "I accept"
                IF (S is preferred to Current-Partner)
                    DUMP2:= Current-Partner
                    Current-Partner:= S
                ELSE
                    DUMP2:= S
                END-IF.
            END-IF.
            move Next-Agent-to-Ask down 1
            offer-outstanding := FALSE
        END-IF.
    CASE: "SEND DUMP MESSAGE X" (X = 1 or 2)
        SEND "I'm no longer interested" to DUMPX
    CASE: "RECEIVE DUMP MESSAGE X" (X = 1 or 2)
        IF RECEIVE "I'm no longer interested"
            Current-Partner:= NIL
        END-IF.
END.
```

Fig. 3. The decentralised version of the modified stable marriage algorithm

The restriction of allowing only single messages to be read or sent in a round raises a number of timing issues. For example, since several agents may make offers to a single agent at one time, an agent may not be able to respond to all its offers in a single round. Consequently, an agent may not get a response to its offer in the same round that the offer was extended. An agent is restricted to having a single offer outstanding at a time.

Agents recognise the end of an offers phase by listening for silence on the communication channels. It is assumed that any agent can detect the presence of any communication, as though the messages were packets being transmitted on Ethernet. If no offers are extended during the EXTEND-OFFERS phase, then each agent is happy with its current position, and the phase can end. Or, an agent might not be able to extend an offer if it is waiting for a response to an offer it has already extended. In this case, the agent sends a dummy message (to itself, if necessary), preventing the phase from ending.

On first examination, the distributed algorithm of Fig. 3 seems to have the problem that an agent may extend an offer, but as soon as the responding agent accepts, the initiating agent dumps the responder. For example, suppose agent A has the preference list B, C, D (from most preferred to least). Agent B's list is E, A. In this case, it would be possible for A to propose to C (having been declined by B in the previous round), and C is willing to accept. As A is offering to C, however, B (having been dumped by E) may offer to A. Agent A accepts the offer from B, and agent C accepts the offer from agent A. When A receives C's response, A's Current-Partner (which was set to B during EXTEND-RESPONSE) is preferred to the responding agent, so A will send a dumping message to C later that same round. This sequence is not problematic, and corresponds to the serial sequence of A offering to C, C accepting, then B offering to A, and A dumping C to join B.

2.6 Unification Phase

At the end of the matching phase, any stable pair (that is, any pairing i, j such that Current-Partner(i) = j and Current-Partner(j) = i) forms a coalition. This two unit entity then enters the next round of negotiations as a single "agent," with a designated head of the coalition who serves as the communication link between the coalition and the other entities (which may be single agents or larger coalitions.) If any coalition formed in the previous round, the algorithm repeats. When no coalition forms, the algorithm terminates.

2.7 Securities Exchange Example

This section traces an example of the coalition formation process applied to a simplified finance domain problem. Agents model brokers, where each agent has a collection of "customer orders," that may be either "buy" orders or "sell" orders for a certain security. Agents exploit the difference in customer perceptions of the values of securities in order to make a profit. Thus, the agents buy from customers that have low sale prices, then resell these securities to other customers that place higher values on them. For example, if agent A had an order from a client who wished to buy 100 shares of XYZ at 50, and agent B had an order to sell 60 shares of XYZ at 40, the pair's profit would be 600 (60 shares transferred, difference of 10 per share). Agents A and B need not split the profit evenly. In fact, if everyone were trying to buy XYZ, but only a few sellers were available, then agent B, the selling agent, would deserve the lion's share of the utility resulting from the sale.

For the communication round, the agents need to exchange two values. The first of these values is the utility that the single agent can earn without forming a coalition, and it is trivial to calculate, by matching as many as possible buy orders to sell orders for each available security, using first the highest buy prices with the lowest sale prices for the same security. For the second piece of information exchanged in the communication phase, an agent needs to tell its coalition partner the full set of customer orders it has, including the number of shares and the price for each order. By combining this information with the receiving agent's knowledge about its own customer orders, each agent can carry out the calculation to determine the value of the coalition. The agent just appends the new orders to its current orders, then proceeds as if it were a single agent with all of the coalition's orders. Since agents outside the coalition cannot affect the coalition's utility in the securities exchange domain (because any two agents can bilaterally execute a trade, outside agents may not "interfere"), the maximum obtainable utility is independent of the excluded agents.

Figure 4 shows that agent 0 has customer orders to buy 100 shares of AAA stock, at any price up to 35; buy 200 shares of BBB as high as 60; buy 100 shares of CCC as high as 50. Other customers have placed sell orders with agent 0, hoping to sell 100 shares of CCC, as low as 45, and 80 shares of DDD as low as 45. The holdings of agents 1 through 3 can be explained similarly. The single agent utility values result from the fact that agents 0 and 2 have buy orders and sell orders for the same stock. That is, agent 0 can buy the 100 shares of CCC from his customer who wants to sell at 45, and then turn around and sell the same shares to his customer willing to buy at 50, thereby keeping the difference of 5 points per share on each of 100 shares.

Agent	BUY	SELL
0	AAA: 100@35 BBB: 200@60 CCC: 100@50	CCC: 100@45 DDD: 80@45
1	DDD: 30@60 EEE: 150@40	BBB: 100@55
2	DDD: 80@35 EEE: 90@55	AAA: 200@30 BBB: 60@50 EEE: 100@50
3	FFF: 75@40 HHH: 100@30	GGG: 200@25

$v(0) = 500$ CCC: 100@5 ($0 \rightarrow 0$)
$v(1) = 0$
$v(2) = 450$ EEE: 90@5 ($2 \rightarrow 2$)
$v(3) = 0$

Fig. 4. Securities exchange example initial holdings and single agent utilities

At the end of the communication round, the agents integrate the information and determine the utility for each of the coalitions which they might join. They then compute the Shapley values of each of the parties in these coalitions, and order the various potential coalition partners based on the expected share of utility that they will obtain. Figure 5 shows the utility to each coalition, and the breakdown of utility calculated by the Shapley value.

$$v(01) = \quad 500 \quad (100\ CCC@5 \quad 0 \to 0)$$
$$ \quad 500 \quad (100\ BBB@5 \quad 1 \to 0)$$
$$ \quad 450 \quad (30\ DDD@15 \quad 0 \to 1)$$
$$ = 1450$$

Shapley Value to $0 = \frac{1}{2}*(500) + \frac{1}{2}*(1450 - 0) = 975$

Shapley Value to $1 = \frac{1}{2}*(0) + \frac{1}{2}*(1450 - 500) = 475$

$$v(02) = \quad 500 \quad (100\ CCC\ @5 \quad 0 \to 0)$$
$$ \quad 450 \quad (90\ EEE\ @5 \quad 2 \to 2)$$
$$ \quad 500 \quad (100\ AAA\ @5 \quad 2 \to 0)$$
$$ \quad 600 \quad (60\ BBB\ @10 \quad 2 \to 0)$$
$$ = \quad 2050$$

Shapley Value to $0 = \frac{1}{2}*(500) + \frac{1}{2}*(2050 - 450) = 1050$

Shapley Value to $2 = \frac{1}{2}*(450) + \frac{1}{2}*(2050 - 500) = 1000$

$$v(03) = \quad 500 \quad (100\ CCC@5 \quad 0 \to 0)$$

Shapley Value to $0 = \frac{1}{2}*(500) + \frac{1}{2}*(500 - 0) = 500$

Shapley Value to $3 = \frac{1}{2}*(0) + \frac{1}{2}*(500 - 500) = 0$

$$v(12) = \quad 1350 \quad (90\ EEE\ @15 \quad 1 \to 2)$$

Shapley Value to $1 = \frac{1}{2}*(0) + \frac{1}{2}*(1350 - 450) = 450$

Shapley Value to $2 = \frac{1}{2}*(450) + \frac{1}{2}*(1350 - 0) = 900$

$$v(13) = \quad 0$$

Shapley Value to $1 = \frac{1}{2}*(0) + \frac{1}{2}*(0 - 0) = 0$

Shapley Value to $3 = \frac{1}{2}*(0) + \frac{1}{2}*(0 - 0) = 0$

$$v(23) = \quad 450 \quad (90\ EEE\ @5 \quad 2 \to 2)$$

Shapley Value to $2 = \frac{1}{2}*(450) + \frac{1}{2}*(450 - 0) = 450$

Shapley Value to $3 = \frac{1}{2}*(0) + \frac{1}{2}*(450 - 450) = 0$

Fig. 5. Utility of two agent coalitions and Shapley values

Figure 5 exhibits two interesting properties of the Shapley value. First, the $v(03)$ shows that including an agent that makes no contribution to the pair (such as agent 3, whose orders are for stocks that none of the other agents can obtain) does not decrease the utility obtained by the other agents in the coalition. Since agent 0 can obtain 500 alone, its group share can not drop below 500. A second interesting point is shown by the coalition of agents 1 and 2. In this case agent 2 is compensated for its option of transferring the 90 shares of EEE stock between its own customers, even though this option is never exercised (i.e., all of the stock that agent 2's customer bought came from agent 1's selling customer.)

Figure 6 gives the preference lists of the four agents, and shows one possible set of formation dynamics for the first round. Since the first step of the algorithm presented in Fig. 2 involves selecting an arbitrary agent from the agent list, there are other dynamics that would work equally well, yielding the same results.

Preference Lists
0: 2, 1
1: 0, 2
2: 0, 1
3:

Agent Selected	Action	Effects
3	3 makes no offers	3 goes solo
1	1 offers 975 to 0	0 accepts
0	0 offers 1000 to 2	2 accepts, 1 dumped
1	1 offers 900 to 2	2 rejects
1	1 makes no offers	1 goes solo
2	2 offers 1050 to 0	0 accepts.

Final Matching
Agent 0 with Agent 2	obtaining 1050
Agent 1 solo	obtaining 0
Agent 2 with Agent 0	obtaining 1000
Agent 3 solo	obtaining 0

Fig. 6. Preference lists and formation dynamics

Once the coalition for the first round has been formed, the process repeats. Agents 0 and 2 now behave like a single entity, and one of them is designated the head of the coalition to serve as the contact for future coalition negotiation. In the second round, agent 1 would be added to the coalition of 0 and 2. A third round would result in no new coalitions, so the process would end, with agents 0, 1, and 2 in a coalition, and 3 relegated to a solo role.

3.   Evaluation

3.1   Computational Complexity

The modified stable marriage algorithm is, like the stable marriage algorithm upon which it is based, a quadratic algorithm. No agent ever approaches the same agent as the initiating agent twice. There are at most $(n-1)^2$ approaches, since in the worst case each agent approaches every other agent. Each invocation of STABLE results in at least one approach, or the removal of one agent from the agent-list, and an agent who voluntarily leaves the agent-list by going solo will never be added back to the agent list. Handling each approach takes only a constant number of operations to process. Therefore, each round of coalition formation is $O(n^2)$. The entire coalition formation process can require at most n rounds, since each round decreases the number of entities by at least one. The whole process is, therefore, $O(n^3)$.

3.2 Modified Algorithm is *not* Stable

As mentioned above, the modified stable marriage algorithm does not maintain the property of stability. There are certain preference orderings for which no algorithm can produce a stable matching, but the algorithm of Fig. 2, or its distributed version in Fig. 3, occasionally leaves unstable pairings even when stable ones exist. The problem can be seen when one agent A is paired to a highly rated choice and therefore rejects the first overture of the second agent, B. If A is dumped, and returns to B, B may be paired with an agent rated more highly than A and reject A. If B is subsequently dumped, it cannot go back to A, since B already asked A, and A refused. Therefore, both A and B are solo when they prefer to be paired with each other.

There are two reasons that instability is to be avoided. The first is that in many cases the unstable solution is not Pareto optimal. Indeed, in the example above, if agents A and B were to form a coalition, they would be better off, without causing any other agents to be worse off. A second reason to avoid instability is that a non-stable solution will probably not be executed as planned. The blocking pair of the non-stable solution would have the incentive to leave their current coalitions and form a new coalition together. Since there is no "enforcement" mechanism that would prevent this defection, it is rational for the two agents to form their own coalition. The algorithms outlined in Fig. 2 and Fig. 3 minimise the ill effects of the instability from the second reason, because an agent is never forced to go solo unless it has been rejected by every agent with which it would like to form a coalition. Therefore, an agent would be less likely to search for ways to break up coalitions since it had already been rejected by every potential partner, so the efforts are likely to be wasted.

3.3 Simulation Results

The centralised approximation of the modified stable matching algorithm was implemented to determine the effectiveness of the matchings that were obtained. In the simulated environment, each pair of agents in the agent pool had a utility that they received if they formed a coalition with each other. The utility values for the coalitions were randomly assigned, from the even numbers in the range of 0 to 998, which the agents split equally. This simulation covered only one "round," so final coalitions were pairs of agents. The results obtained by the modified stable marriage approach are compared to those obtained by exhaustively generating all of the possible pairings and choosing the matching that yielded the highest total utility to the community. There was no restriction that the matching chosen by the exhaustive search must be stable.

The results are shown in Fig. 7. The x-axis indicates the number of agents that were involved in the negotiation. System memory limitations caused the naive exhaustive search algorithm to fail on problems with more than 12 agents. The y-axis shows the amount of utility that the average agent obtained, averaged over 100 runs. The maximum value that any agent could achieve is 499. Comparing the results of the exhaustive algorithm which examined all of the possible pairings with the modified stable marriage algorithm which examines only a subset of the search space shows, as expected, that the stable marriage approach does not do as well in absolute terms. The modified stable marriage algorithm does obtain a substantial fraction of the optimal utility, however, as Table 1 shows. The "Efficiency" column shows the utility obtained by the stable algorithm as a percentage of the optimal values obtained for the same data.

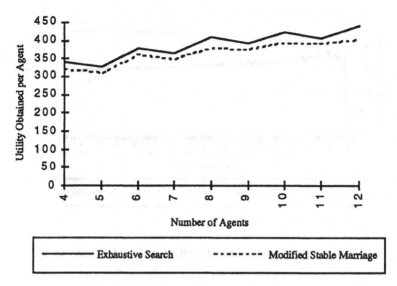

Fig. 7. Performance compared to optimal

Table 1. Efficiency compared to optimal

# of Agents	Optimal	Stable	Efficiency
4	340.3	319.5	0.939
5	326.6	309.0	0.946
6	377.4	358.6	0.950
7	363.3	347.1	0.956
8	406.6	379.3	0.933
9	388.7	373.9	0.962
10	422.7	391.9	0.927
11	404.2	393.7	0.974
12	437.5	401.6	0.918

As the size of the agent pool increases, the trade-off between computational effort and quality of the resulting matching favours the modified stable marriage approach, since exhaustive search is growing exponentially, but the modified stable marriage is growing only quadratically. Figure 8 shows that the stable marriage approach retains its ability to find high quality solutions even as the problem size grows very large. In fact, as the number of agents increases, the utility that the average agent obtains nears the absolute maximum of 499, which is not necessarily reachable in every problem instance. Moreover, run times show only a modest increase as the size of the agent set increases, and even sets as large as 350 agents could be handled in less than 1 second using the modified stable marriage algorithm.

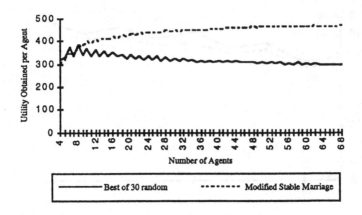

F ig . 8. Performance compared to "Best of 30 random matchings"

Figure 8 compares the stable marriage approach with the simple approach of picking several different pairings at random, and then choosing the one of that set which yields the highest utility for the community. These results are averaged over 100 runs, where each run included selecting 30 different random pairings. The random algorithm took approximately as long to evaluate these 30 different options as the stable marriage method did to come up with its solution for the largest problem sizes. The random approach performed well at small problem sizes, which is unsurprising since there are fewer than 30 different matchings with 4 agents, so the best-of-30 approach is equivalent to an exhaustive search. As the problem size increased, however, the random approach did not perform as well, while the stable marriage algorithm improved noticeably.

The saw-toothed nature of these graphs is due to the coalitions of pairs. Where the number of agents is odd, one agent is forced to go solo, and receives no utility. The "odd one out" pulls down the average by increasing the number of agents without increasing the utility.

4. Conclusions

4.1 Summary

The algorithm presented here for coalition formation uses game theory to address a problem that has not received much attention in DAI. Since there are cases where agents can bilaterally form coalitions and it is rational for them to do so, it is necessary to plan for this process of coalition formation and perhaps even encourage it as a viable scheme of organising agents. The proposed variant of the stable marriage algorithm creates an efficient environment for the search through possible coalitions, and results in the formation of coalitions that are close to the theoretical optimal under the given conditions. The Shapley value is one possible fairness criterion that is easy to compute (for the two agent case), and removes the need for further negotiation about the division of the coalition's utility.

4.2 Areas for Future Work

By combining portions of Irving's stable roommate algorithm with the approach suggested here, it may be possible to unite the strong points of both methods and design a fully distributed algorithm that finds a stable match whenever one exists. Further work could also be done to handle the case where no stable matching exists. Although the Shapley value is fair for determining the contribution of the two entities that make up the coalition, the division of utility within a single entity may require a different mechanism. Also, the calculation of the Shapley value used here assumes that the two agents agree on $v(AB)$, and that this value is, in fact, the correct one. The calculation of a fair value is not as easy if these assumptions do not hold. For one possible solution to these last two problems, see [8].

Underlying the application of the stable marriage problem to the coalition formation problem is the assumption that coalitions may form by merging two groups at a time. If the utility is a non-decreasing function of the number of agents, this strategy will allow the proper coalitions to form. The securities exchange domain meets this condition, since adding a new agent to a coalition never drags down the utility of the group. However, by restricting negotiations to only bilateral negotiations, many cases of beneficial collaboration may be missed. For example, consider the three agent exercise where the group of three obtains a large utility reward, but any pair of agents has a negative group utility value. In this instance, there will be no incentive for any two agents to get together, so all collaboration will be blocked, and the potential utility gain from the grand coalition will go unrealised. Analysis should be carried out to determine what classes of domains meet this criterion.

4.3 Acknowledgements

This work would not have been possible without the guidance and support of the numerous people who made considerable contributions. Steven Brenner, Barbara Grosz, Victor Milenkovic, Jeff Rosenschein, Dan Roth, and Stuart Shieber deserve mention for thought-provoking discussions and providing valuable references.

References

1. Davis, R., Smith, R.G.: Negotiation as a metaphor for distributed problem solving. Artificial Intelligence **20** (1983) 63-109
2. Ephrati, E., Rosenschein, J.S.: The Clarke Tax as a consensus mechanism among automated agents. In: Proceedings of the Ninth National Conference on Artificial Intelligence. The AAAI Press, Cambridge MA, 1991, pp. 173–178
3. Feldman, J., Sproull, R.: Decision Theory and Artificial Intelligence II: The Hungry Monkey. In: Allen, J.F., Hendler, J., Tate, A. (eds.) Readings in Planning. Morgan Kaufmann, San Mateo CA, 1990, pp. 207–224
4. Grosz, B., Sidner, C.L.: Plans for discourse. In: Cohen, P.R., Morgan, J.L., Pollack, M.E. (eds.) Intentions in Communication. Bradford Books at MIT Press, Cambridge MA, 1990, pp. 417–444
5. Grosz, B., Kraus, S.: Collaborative plans for group activities. In: Proceedings of the Thirteenth International Joint Conference on Artificial Intelligence. Morgan Kaufmann, San Mateo CA, 1993, pp. 367–375
6. Gusfield, D., Irving, R.W.: The Stable Marriage Problem: Structure and Algorithms. MIT Press, Cambridge MA, 1989
7. Ketchpel, S.P.: Deal Making Among Agents of Different Abilities. Unpublished manuscript, 1991
8. Ketchpel, S.P.: Forming Coalitions in the Face of Uncertain Rewards. In: Proceedings of the Twelfth National Conference on Artificial Intelligence. To appear.
9. Kraus, S., Wilkenfeld, J., Zlotkin, G.: Multiagent Negotiation Under Time Constraints. CS-TR-2975, University of Maryland, 1992.

10. Lochbaum, K.E., Grosz, B., Sidner, C.L.: Models of plans to support communication: An initial report. In: Proceedings of the Eighth National Conference on Artificial Intelligence. The AAAI Press, Cambridge MA, 1990, pp. 485–490
11. Raiffa, H.: The Art and Science of Negotiation. Belknap Press of Harvard University Press, Cambridge MA, 1982
12. Rapoport, A.: N-person game theory; concepts and applications. University of Michigan Press, Ann Arbor MI, 1970
13. Rosenschein, J.S.: Personal communication, 1993
14. Wellman, M.: A general-equilibrium approach to distributed transportation planning. In Proceedings of the Tenth National Conference on Artificial Intelligence. The AAAI Press, Cambridge MA, 1992, pp. 282–289

Organizational Fluidity and Sustainable Cooperation

Natalie S. Glance and Bernardo A. Huberman

Dynamics of Computation Group
Xerox Palo Alto Research Center
Palo Alto, CA 94304
{glance, huberman}@parc.xerox.com

Abstract: We show that fluid organizations display higher levels of cooperation than attainable by groups with either a fixed social structure or lacking one altogether. By moving within the organization, individuals cause restructurings that facilitate cooperation. Computer experiments simulating fluid organizations faced with a social dilemma reveal a myriad of complex cooperative behaviors that result from the interplay between individual strategies and structural changes. Significantly, fluid organizations can display long cycles of sustained cooperation interrupted by short bursts of defection.

1 Introduction

The study of social dilemmas provides insight into a central issue of social behavior: how global cooperation among individuals confronted with conflicting choices can be secured. In recent work, we have shown that cooperative behavior in a social setting can be spontaneously generated, provided that the groups are small and diverse in composition, and that their constituents have long outlooks [1, 2]. Furthermore, the emergence of cooperation takes place in an unexpected fashion, appearing suddenly and unpredictably after a long period of stasis.

In our model of ongoing collective action, intentional agents make choices that depend on their individual preferences, expectations and beliefs as well as upon incomplete knowledge of the past. Because of the nature of individuals' expectations, a strategy of conditional cooperation was shown to emerge from individually rational choices. According to this strategy, an agent will cooperate when the fraction of the group perceived as cooperating exceeds a critical threshold.

Since the critical threshold grows with group size, beyond a certain size, cooperation is no longer possible in flat, structureless groups [1, 3]. However, groups are often characterized by a distinct social structure that emerges from the pattern of interdependencies among individuals. Clustering in a social structure effectively decreases the size of the group as each individual cares most about the behavior of his/her own cluster. Accordingly, we will show how cooperation is a more likely outcome for hierarchically structured groups. In addition, isolated clusters of cooperation can survive and even trigger a cascade of cooperation throughout the rest of the organization.

The potential for cooperative solutions to social dilemmas increases further if groups allow for structural changes. In a fluid structure, the pattern of interdependencies can vary widely over time, since the sum of small local changes in the structure of a group results in broad restructurings. We find that fluid groups show much higher levels of cooperation over time. By moving within the group structure, individuals cause restructurings that enable cooperation. Computer experiments simulating fluid organizations faced with a social dilemma reveal a myriad of complex behaviors that result from the interplay between various amounts of cooperation and structural changes.

The advantages of fluidity must be balanced against possible losses of effectiveness for an organization. By effectiveness, we mean how productive a given organization is in obtaining an overall utility over time. The production function for a social good will depend on the tightness or looseness of clustering, and possibly on the stability of the social structure. The form of the production function is not known *a priori*; however, given an optimal level of clustering for a particular good, we illustrate that there is an ideal range of fluidity associated with it.

Although our results are predicated upon the assumption of hierarchical structures, we expect them to generalize to less constrained forms of social networks. Any types of structure which exhibit clustering should qualitatively behave like the hierarchies studied herein. Whether or not this is the case, the effect of the structure and fluidity of social networks on social behavior remains an interesting theoretical avenue to explore, both within and outside the study of social dilemmas.

2 The Topology of Organizations

Organizational theorists study group structure in order to elucidate the nature and working of the firm. In firms, there is generally an informal structure that emerges from the pattern of affective ties among participants as well as a formal one imposed from above [4]. In our study, we will be interested in the former type, the network of activities and interactions that link individuals in a group, commonly known as the sociometric structure.

A school of sociologists and social psychologists also regard the social structure of a group as elemental in understanding group behavior. The approach called social network analysis began with Barnes' [5] and Bott's [6] first attempts to use the relationship of the linkages in a network to interpret social action, and has gained momentum in the previous decade [7]. From the micro view of individual interdependencies emerges a global perspective on social structures.

Thus, organizational structure in social groups can spontaneously emerge from the pattern of interactions of group members. In contrast with its typical usage in sociology, we use the notion of group in a loose sense, defining it as the community affected by a particular problem or situation. There are a number of different general types of structural topologies: flat, or structureless, hierarchical, matrix, circular, linear, and many others. In this paper we restrict ourselves to the first two.

In addition to its topology, an organizational structure can also be described by the amount of fluidity it exhibits. Fluidity encompasses such features as how flexible the structure is, how readily individuals can locally modify structure by changing the strength of their interaction with others. The concept of fluidity also has a precedent in the social sciences; for example Srinivas and Béteille [8] state that "a network even when viewed from the standpoint of a single individual has a dynamic character. New relations are forged, and old ones discarded or modified."

Consider, as a concrete example, the social problem of limiting air pollution. This is a problem that on one level the whole world faces and must solve collectively. The common good is clean air with a minimum of pollution, not only to make living conditions better today, but, more importantly, to ensure that the world remain hospitable to life in the future. Clean air is a common good because everyone benefits from it independently of other individual's efforts to limit pollution.

In this example, the impact of pollution depends partially on relative geographical locations. That is, neglecting prevailing winds and currents, a person is more bothered when her neighbor burns his compost pile than when someone across town does the same. Similarly, the cumulative effect of everyone in town driving their cars to the popular bookstore (instead of bicycling) affects a local resident much more than someone else who lives miles away. Of course, some individuals' spheres of influence will range much further than others'. This dilution of impact with distance can be represented as a hierarchy of interactions which reveals itself in the unraveling of layer upon layer: neighborhood, town, county, state, world. The effect of one individual's actions on another depends on how many layers apart they are.

Within this hierarchical structure there is some fluidity. Although people are constrained by their resources and personal ties, they can often choose to move to a new location to escape a neighbor's radio or a textile company's fumes.

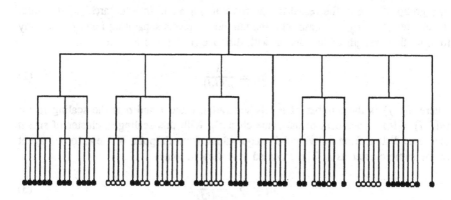

Fig. 1. A hierarchical organization can be visualized as a tree. The tree reveals the structure of the whole: each branching represents a subdivision of the higher level. The nodes at the lowest level represent individuals, filled circles mark cooperators, open circles, defectors. One can read directly from the tree the number of layers of the organization separating any two individuals by simply counting nodes backwards up the tree from each individual until a common ancestor is reached. This number determines the distance between two individuals. The larger the distance between them, the less their actions will affect each other.

We now quantify the notion of many levels of clustering within a group. For a group with a hierarchy of connection strengths among the members, the organizational structure can be represented graphically by a tree, as in Fig. 1. The technique of representing the interdependencies in social networks by hierarchical clustering in tree-like structures dates from the work of Hartigan [9], and more recently, Burt [10]. The level of interaction between two individuals can be read directly from the tree. Tightly clustered individuals share an ancestor node one level up in the tree. Those more loosely coupled may be connected two levels up. In general, interaction strength is indicated by the number of levels to a common ancestor node. Thus the tree gives a visual description of the amount and extent of clustering in the group. Note that this interpretation of hierarchy as a pattern of interdependencies is divorced from any notion of rank within the group.

A second more general way to describe the pattern of interactions among individual members of a group is to define the matrix of interdependencies between them. Define A to be the matrix of interactions; then a_{ij} is the strength of interaction between individual i and j. This formalism permits generalization to many other topologies of group structure apart from hierarchical ones.

From this formalism it is easy to determine the extent of influence on any particular individual by the rest of the group. We call the cumulative influence perceived by individual i the rescaled quantity

$$\tilde{n}_i = \sum_j a_{ij}. \tag{1}$$

In a flat group, one in which all influences are equal, A would be filled with 1's, and in a group of size n, the rescaled size $\tilde{n}_i = n$, for all i. In a hierarchical structure, the components, a_{ij}, decrease with the number of levels separating i and j. One way to scale the strength of interaction with distance in the tree is to let

$$a_{ij} = \frac{1}{a^{d(i,j)}} \tag{2}$$

where $d(i,j)$ is the number of levels separating i and j and a is the scaling factor ($d(i,j) = 0$ for members of the same cluster). With this scaling, a cluster of size a one level distant from individual i is equivalent, from i's point of view, to one agent in the same cluster as i. The rescaled size then becomes

$$\tilde{n}_i = \sum_j \frac{1}{a^{d(i,j)}}. \tag{3}$$

Along with social structure comes the notion of fluidity. In a fixed structure the pattern of interactions remains fixed in time. This need not be the case. In the example tying relative geographical positions to influence, the pattern of interactions changes slightly every time someone moves to a new locale, and more significantly when whole groups dissolve and reform. In a corporation, for example, employees may have some leeway to switch departments or even to move between regional branches. More importantly, the sum of small local changes in the structure of a group can result in broad restructurings over time.

The ease with which such moves can be accomplished reveals the amount of fluidity in the group structure. The notion of fluidity actually consists of two elements.

One is how easily individuals can move within the structure, the second, how easily they can break away on their own, thereby extending the structure. These moves are restructurings in the sense that they change the pattern of interactions. Breaking away can range from moving to a secluded area, to branching off to start a new work group, to founding a new company. In many structured groups there will be costs associated with both moving within the structure or breaking away. The higher the costs the less fluid the structure will be, as individuals must anticipate a clear advantage to themselves before moving or breaking away.

3 Social Dilemmas

There is a long history of interest in collective action problems in political science sociology, and economics [11, 12]. Hardin coined the phrase "the tragedy of the commons" to reflect the fate of the human species if it fails to successfully resolve the social dilemma of limiting population growth [13]. Furthermore, Olson argued that the logic of collective action implies that only small groups can successfully provide themselves with a common good [14]. Others, from Smith [15] to Taylor [16, 17], have taken the problem of social dilemmas as central to the justification of the existence of the state. In economics and sociology, the study of social dilemmas sheds light on, for example, the adoption of new technologies [18] and the mobilization of political movements [19].

In a general social dilemma, a group of people attempts to obtain a common good in the absence of central authority. Each individual has two choices: either to contribute to the common good, or to shirk and free ride on the work of others. The payoffs are structured so that the incentives in the game mirror those present in social dilemmas. All individuals share equally in the common good, regardless of their actions. However, each person that cooperates increases the amount of the common good by a fixed amount, but receives only a fraction of that amount in return. Since the cost of cooperating is greater than the marginal benefit, the individual defects. Now the dilemma rears its ugly head: each individual faces the same choice; thus all defect and the common good is not produced at all. The individually rational strategy of weighing costs against benefits results in an inferior outcome — no common good is produced.

However, the logic behind the decision to cooperate or not changes when the interaction is ongoing since future expected utility gains will join present ones in influencing the rational individual's decision. In particular, individual expectations concerning the future evolution of the game can play a significant role in each member's decisions. The importance given the future depends on how long the individuals expect the interaction to last. If they expect the game to end soon, then, rationally, future expected returns should be discounted heavily with respect to known immediate returns. On the other hand, if the interaction is likely to continue for a long time, then members may be wise to discount the future only slightly and make choices that maximize their returns on the long run. Notice that making present choices that depend on the future is rational only if, and to the extent that, a member believes its choices influence the decisions others make.

One may then ask the following questions about situations of this kind: if agents make decisions on whether or not to cooperate on the basis of imperfect information about group activity, and incorporate expectations on how their decision will affect other agents, then how will the evolution of cooperation proceed? In particular, how will the structure and fluidity of a group affect the dynamics?

In [1] we borrowed methods from statistical thermodynamics [20], a branch of physics in order to study the evolution of social cooperation. This field attempts to derive the macroscopic properties of matter (such as liquid versus solid, metal or insulator) from knowledge of the underlying interactions among the constituent atoms and molecules. In the context of social dilemmas, we adapted this methodology to study the aggregate behavior of a group composed of intentional individuals confronted with social choices. This allowed us to apply results from theoretical physics to the study of the dynamics of group cooperation.

In our mathematical treatment of the collective action problem we stated the benefits and costs to the individual associated with the two actions of cooperation and defection, *i.e.* contributing or not to the social good. The problem thus posed is referred to in the literature as the n-person prisoner's dilemma [21, 16, 3]. We also allowed beliefs and expectations about other individuals' actions in the future to influence each member's perception of which action, cooperation or defection, will benefit it most in the long run. Using these preferences functions, we applied the stability function formalism [22] to provide an understanding of the dynamics of cooperation in the case of groups with no organizational structure. We concluded that the emergence of cooperation among individuals can take place in a sudden and unexpected fashion. This finding turns out to be useful when examining the possibility for cooperation in fluid organizations.

4 Structures for Cooperation

We showed in our earlier work that there is a clear upper limit, n^*, to the size of a group which can support cooperation. Here we examine how this limit can be stretched by a hierarchically structured group whose members are able to move freely within the group. (For the mathematical background behind the dynamics of cooperation in structured groups, the reader is referred to the longer version of this paper [23].) First, we present the limitations of flat groups and groups with fixed structures. Then we show to what extent these limitations can be overcome by fluid groups. Finally, we discuss the possible trade-off between fluidity and effectiveness in organizations.

4.1 Flat Groups

Results of Monte Carlo simulations conducted in asynchronous fashion confirm the theoretical predictions for structureless groups obtained using the stability function formalism of [22, 1]. Each individual decides whether to cooperate or defect based on the criterion that the perceived fraction cooperating, f^c, must be greater than a critical fraction, f^{crit}. Uncertainty enters since these decisions are based on perceived levels of cooperation which differ from the actual attempted amount of cooperation.

As shown in the previous work, there is a critical group size beyond which cooperation will not be sustained. There is also a range of group sizes below the critical size for which the system has two equilibrium points, one with most of the group defecting, the other with most members cooperating.

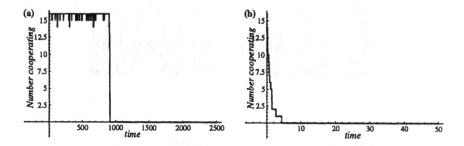

Fig. 2. The evolution of cooperation is shown here for a structureless group of size $n = 16$. The benefit for cooperating, b, is 2.5, the cost, c, is 1, the probability p that agent cooperates successfully is 0.95, the reevaluation rate α is 1, and the delay τ is also 1. (a) Given these parameters, the horizon length is set at 12.0 to provide an example in which cooperation is the metastable equilibrium and defection, the global equilibrium. Indeed, the group cooperates for a very long time, over 900 time steps. The outbreak of defection is sudden and unpredictable, occurring at widely different times in numerous simulations of the same system. (b) Here the horizon length is set to 8.9; defection becomes the only equilibrium state for the system. As predicted by the theory, the group relaxes exponentially fast to the defecting state.

In Fig. 2(a), the evolution of the system is shown for a group of size 16 whose horizon length is such that the critical size for cooperation is less than 16. The system begins in the metastable overall cooperation state and remains there for a very long time. Defection is the global equilibrium, however, and eventually, the transition to defection occurs. In (b), the horizon length is such that the critical size is greater than 16. In this case, the only equilibrium point is overall defection; accordingly, the onset of defection is swift for a group initially cooperating.

4.2 Fixed Structures

Imagine now that the pattern of interactions among the 16 individuals depicts a group broken down into 4 clusters of 4 agents each. If the amount of influence one agent has on another scales down by a factor of $a=4$ for each level in the hierarchy separating two individuals, then the rescaled size of the group is $\tilde{n} = 4 + 12/4 = 7$ for all the agents (symmetric structure). If only because rescaled group size is smaller, cooperation can be sustained under more severe conditions for a hierarchical group than for a flat one.

More interestingly, an enclave of cooperation within the hierarchy can initiate a widespread transition to cooperation within the entire organization. The mechanism that explains this cascade to cooperation results from the clustering of agents and

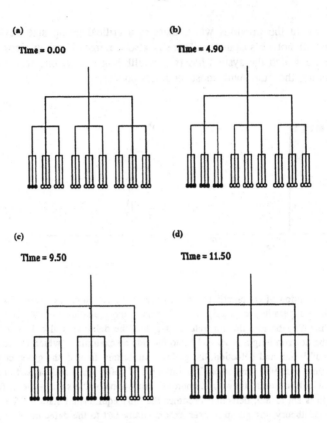

Fig. 3. Overall cooperation in a hierarchically structured group can be initiated by the actions of a few agents clustered together. These agents reinforce each other and at the same time can spur agents one level further removed from them to begin cooperating. In turn, this increase in cooperation can spur cooperation in agents even further removed in the structure. In this example, the structure is fixed: the three level hierarchy consists of three large clusters, each subsuming three clusters of three agents each. The parameters in this example were set to $H = 10.0$, $b = 2.5$, $c = 1$, $p = 0.97$, $\alpha = 1$, and $r = 0$.

cannot occur in a flat group. The cascading phenomenon is more vivid in hierarchies with several layers; consider for example a group of 27 agents structured as in Fig. 3(a). In this case, the rescaled size $\bar{n} = 3 + 6/3 + 18/3^2 = 7$ for all the agents. Filled circles at the lowest level represent cooperating agents, open circles are defecting agents. Individuals in different parts of the organization will observe widely differing amounts of cooperation. Those in the leftmost cluster see a fraction of about 3/7 cooperating while agents one level away see 1/7 cooperating and agents two levels away see a fraction 0.33/7 cooperating. So if $1/7 \lesssim f^{crit} \lesssim 3/7$, the agents in the leftmost cluster can sustain cooperation almost indefinitely among the three of them despite the fact that the rest of the organization is defecting. On the other hand, although agents one level away may be defecting initially, for certain parameter choices, cooperation may actually be the long-term stable state for those clusters in the presence of uncertainty. If these agents in clusters one level away start cooperating, they may then trigger the onset of cooperation in the clusters two levels away from the initial cooperators. Fig. 3 shows four snapshots in time of a group of 27 agents exhibiting this sequence of behavior over time.

However, as with flat groups, groups with fixed structures easily grow beyond the bounds within which cooperation is sustainable. In the case mentioned earlier of 16 individuals broken down into four clusters, when the rescaled size is larger than $\tilde{n}^* = 4.42$, the group rapidly evolves into its equilibrium state of overall defection. But even when the rescaled size is less than $\tilde{n}^*$, cooperation will be metastable in the sense that the agents may remain cooperating for a very long time but eventually a transition to overall defection will occur. When the transition finally happens, the onset of defection is observed to be very rapid, as for flat groups.

4.3 Fluid Structures

In fluid structures, individual agents are able "move" within the organization. The way in which this might occur depends on what determines the structure, be it geographical location in the world or type of work within a company. The amount of fluidity in an organization is variable. It depends on two factors: (1) the ease with which an individual can switch between two clusters; and (2) how readily agents break away to form clusters of their own. Globally, the sum of individual moves between clusters translates into a mixing and merging of the agents. On the other hand, the local decisions made by individuals to break away and start new clusters expands the structure by decreasing the extent of clustering.

In our simulations, agents regularly reevaluate the situation. Previously, they had only one choice to make: whether to cooperate or defect. In fluid organizations, they must also evaluate how satisfied they are with their location within the structure. However, the agents are assumed to evaluate only one of these two choices at any given decision point.

Individuals make their decision to cooperate or defect according to the long-term benefit they expect to obtain, as before. In order to evaluate their position in the structure, an individual compares the long-term payoff it expects if it stays put with the long-term payoff it expects if it moves to another location, chosen randomly. In these calculations, the agent's strategy in response to the social dilemma remains the same, be it cooperation or defection. In order to determine the payoff it expects to obtain by moving, the individual must have access to the world as seen by the individual whose position it is evaluating. This additional information is not required by individuals in groups whose structure is fixed. Thus, the validity of this model of fluid organizations is limited by the extent to which this information is available.

In addition, there might be a barrier to moving, either because there is a cost associated with it, or because the agents are risk-averse. So for example, an individual might move only if it perceives the move to increase its expected payoff by a certain percentage, which we shall refer to as the moving barrier.

When evaluating its position, the individual also considers the possibility of breaking away to form a cluster of its own. The agent will do so if it perceives the

Time = 0.00   Time = 2.20   Time = 43.20

Time = 62.20   Time = 58.40

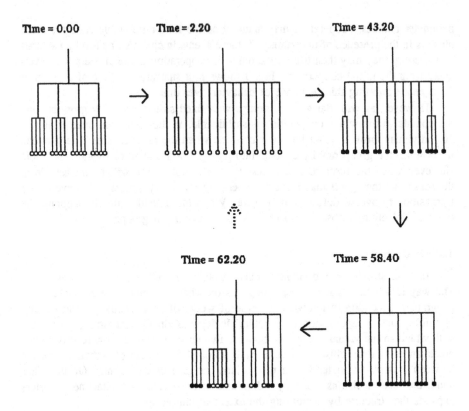

Fig. 4. This figure highlights the ability of fluid groups to recover from overall defection among its members. Initially all members of the group, divided into four clusters of four agents each, are defecting. The next snapshot in time shows that almost all of the agents have broken away on their own. In this dispersed structure, agents are much more likely to switch to a cooperative strategy, and indeed, by the next snapshot shown, all individuals are cooperating. Because of uncertainty, however, agents will occasionally switch between clusters. Eventually, a cluster grows large enough that a transition to defection begins within that cluster. At the same time, individuals will be moving away to escape the defectors. We see these processes happening in the fifth snapshot. At this point, more and more agents will break away on their own, and a similar cycle begins again. The parameters in this example were set to $H = 4.75$, $b = 2.5$, $c = 1$, $p = 0.9$, $\alpha = 1$, and $r = 0$, as in the previous figure.

payoff for either staying put or moving to be small enough that it feels it has nothing to lose by taking a chance and starting its own cluster. The agent can only break away one level at a time, so from a cluster of several agents, it may break away to form a cluster on its own one level distant from its parent cluster. The next time this same individual reevaluates its position, it can then break away an additional layer distant from its parent cluster, if no other agent has come to join it in its new cluster. In this way an agent can break away many levels from its original cluster.

How easily agents are tempted to break away determines the break away threshold. We will give these thresholds as a fraction of the maximum possible payoff over time. Higher thresholds indicate that an individual is more likely to be unsatisfied with both its present position and the alternatives and thus will tend to break away.

Computer experiments implementing this notion of fluid organizations reveal a myriad of complex behavior. Through local moves and break aways, the organization can often restructure itself to recover either from outbreaks of defection or to overcome an initial bias to defection in the group. A series of snapshots over time for a group of size 16 are shown in Fig. 4. Initially, the group is divided into four clusters of four defecting agents each — this represents an extreme condition and is thus a good test of the group's ability to overcome defection. A later snapshot shows that most of the agents have broken away on their own. This restructuring enables the global switchover to cooperation since the rescaled size is now small enough that cooperation is the global equilibrium for the system, were the structure to remain fixed. Fixed it is not, however, and over time, the now mostly cooperating agents cluster back together again, moving towards clusters where they perceive the amount of cooperation to be highest. Eventually, one or more clusters will grow too large to support cooperation indefinitely, creating the potential for an outbreak of defection. Each outbreak is quelled by a process similar to that the responsible for the initial recovery. Cycles of this type have been observed to appear frequently in the lifetime of the simulated organization.

4.4 Effectiveness *vs.* Fluidity

In a very fluid organization, individuals break away often, thus founding new clusters, and move between clusters very readily. On the other hand, these actions rarely occur in a group with little fluidity in its structure. In the example of a fluid organization given above, the amount of fluidity of movement was set to an intermediate amount: the moving barrier was set at 15% and the break away threshold was 45% of the maximum possible payoff.

Observations of many simulations at various levels of fluidity and for differing values of horizon length point to the following conclusions. High break away thresholds mean that individuals tolerate little deviation from the maximum available payoff and often break away in search of "greener pastures." Since breaking away is the primary mechanism that allows a group to recover from bouts of defection, a greater tendency to break away is favorable in this sense. However, higher break away thresholds also cause the structure to become very dilute and disconnected. In the extreme case, agents will tend to always want to be on their own, as in the second snapshot of Fig 4. On the other hand, large moving barriers inhibit the clustering of agents and cause the structure to vary little over time. Thus large moving barriers help stabilize the structure. If the group is cooperating, its structure may then remain relatively fixed over time. Coupled with high break away thresholds, large moving barriers mean that a cooperating system may be frozen into a structure with a very small amount of clustering.

In general, the combination of ease of breaking away and difficulty of moving between clusters enable the highest levels of cooperation. However, this yields the counter-intuitive result of cooperating dis-organizations! That is, agents are all cooperating, but on their own.

The reason this seems counter-intuitive is because, thus far, the effectiveness of the organization has not been considered. By effectiveness, we mean how productive a given organization is in obtaining an overall utility over time. This is revealed by how well an organization achieves its goals. It seems reasonable to assume that each organization operates most effectively when its structure exhibits a certain amount of clustering. The ideal amount of clustering for a particular organization will depend on type of good the group is attempting to provide itself with and how this good is produced. Determining the optimal amount of clustering for a given type of organization is beyond the scope of this paper; however, we can say something about the range of fluidity that allows an organization to be most effective given an optimal level.

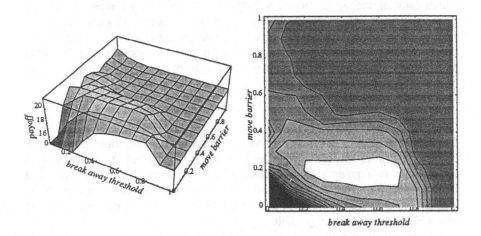

Fig. 5. In (a) the total actual payoff to the group, averaged over thousands of time steps, is plotted as a function of the break away threshold and move barrier. High break away thresholds and low move barriers correspond to high amounts of fluidity in the structure of the organization. The contour plot is given in (b) to highlight the gradient of the functional dependence of payoff on fluidity. In this example, the optimal amount of clustering is taken to be $\tilde{n}^{clust} = 10$, a level corresponding to the clustering of the group into two large subgroups. These results show that there is a range in the amount of fluidity at which this organization operates most effectively. The other parameters values in this example are $H = 8$, $b = 2.5$, $c = 1$, $p = 0.95$, $\alpha = 1$, and $r = 0$.

To factor in effectiveness, we must modify the production function of the common good to reflect increased performance at optimal clustering levels. One way to do this is to posit that the benefit to the group for a cooperative action depends on the amount of clustering. To this end, we introduce a variable

$$\tilde{n}^{clust} = \frac{1}{n} \sum_i \tilde{n}_i \qquad (4)$$

that indicates the average amount of clustering within the structure and postulate that the benefits of cooperation are highest when the amount of clustering is equal to the optimal amount $\tilde{n}^{opt}$ and falls off to either side. Qualitatively, then, the benefit of an individual's cooperative action looks like

$$b' = b \exp\left[-\left(\tilde{n}^{opt} - \tilde{n}^{clust}\right)^2 / n^2\right], \tag{5}$$

where b is the benefit in a flat group.

The interplay between fluidity and effectiveness is best observed when the critical rescaled size falls inside a certain regime. Within this regime, the critical size is such that high levels of clustering cause the group to be unstable to defection while low levels allow to group to recover. To measure effectiveness, we keep track of the average actual payoff over time for varying amounts of fluidity, given a fixed optimal average amount of clustering. Once again, we consider a group of 16 agents, with a hierarchical structure two levels deep. The ideal amount of clustering is set at $\tilde{n}^{clust} = 10$. Such a high level of clustering occurs when the agents break up into two large clusters. Fig. 5 shows that there is indeed a range in the amount of fluidity at which the organization operates most effectively. Systems operating within this range of fluidity may cooperate less over time than those with less fluidity, but compensate by having higher levels of clustering.

5 Discussion

We have shown in this paper that fluid organizations display higher levels of cooperation than attainable by groups with either a fixed social structure or lacking one altogether. In fluid organizations, individuals can easily move between groups or even strike out on their own. The sum of many such moves results in a global restructuring of the organization over time.

In an organization faced with the social dilemma of providing itself with a collective good, such ongoing incremental restructurings can provide free riders with the incentive to cooperate. That is, in a group with many free riders, individual moves will cause the structure to disperse into many small clusters. Cooperation is much more likely to emerge spontaneously within these small clusters and then spread to the rest of the organization.

The results of extensive computer experiments bear out these conclusions. In particular, fluid organizations can display long cycles of sustained cooperation interrupted by short bursts of defection. The average level of cooperation sustained over time depends on the amount of fluidity in the organization as well its breadth and extent. A point to consider, however, is that the advantages of fluidity must be balanced against a possible accompanying decrease in the effectiveness of the group.

Although our model assumes hierarchical groups faced with a social dilemma, we expect our results will generalize to less constrained forms of social networks and to more general production problems. Thus, fluidity may play a crucial role in ensuring that a large organization attain high levels of cooperation. In fact, modern firms have begun to foster organizational fluidity: a central theme in recent reorganizations of large corporations has been the creation of pathways that allow project-centered groups to rapidly form and reconfigure as circumstances demand. A recent cover story in *Fortune* magazine claims that "corporations are finally realizing the need to recognize the informal organization, free it up, and provide it the resources it needs," citing such corporate powerhouses as Apple Computer, Levi Strauss, and Xerox [24].

References

[1]   Natalie S. Glance and Bernardo A. Huberman. The outbreak of cooperation. *Journal of Mathematical Sociology*, 17(4):281–302, 1993.

[2]   Bernardo A. Huberman and Natalie S. Glance. Diversity and collective action. In H. Haken and A. Mikhailov, editors, *Interdisciplinary Approaches to Complex Nonlinear Phenomena*, pages 44–64. Springer, New York, 1993.

[3]   Jonathan Bendor and Dilip Mookherjee. Institutional stucture and the logic of ongoing collective action. *American Political Science Review*, 81(1):129–154, 1987.

[4]   W. Richard Scott. *Organizations: Rational, Natural, and Open Systems*. Prentice-Hall, Inc., Englewood Cliffs, New Jersey, 1992.

[5]   J. A. Barnes. Class and committees in a norwegian island parish. *Human Relations*, 7:39–58, 1954.

[6]   Elizabeth Bott. Urban families: conjugal roles and social networks. *Human Relations*, 8:345–383, 1955.

[7]   C. P. M. Knipscheer and Toni C. Antonucci, editors. *Social Network Research: Substantive Issues and Methodological Questions*. Swets and Zeitlinger, Amsterdam, 1990.

[8]   M. M. Srinivas and A. Beteille. Networks in indian social structure. *Man*, 64:165–168, 1964.

[9]   J. A. Hartigan. Representation of similarity matrices by trees. *Journal of the American Statistical Association*, 62:1140–1158, 1967.

[10]   R. S. Burt. Models of network structure. *Annual Review of Sociology*, 6:79–141, 1980.

[11]   Thomas C. Schelling. *Micromotives and Macrobehavior*. W. W. Norton and Company, Inc., United States, 1978.

[12]   Russell Hardin. *Collective Action*. Johns Hopkins University Press, Baltimore, 1982.

[13]   Garrett Hardin. The tragedy of the commons. *Science*, 162:1243–1248, 1968.

[14]   Mancur Olson. *The Logic of Collective Action*. Harvard University Press, Cambridge, 1965.

[15]   Adam Smith. *The Wealth of Nations*. Random House, New York, 1937.

[16]   Michael Taylor. *Anarchy and Cooperation*. John Wiley and Sons, New York, 1976.

[17]   Michael Taylor. *The Possibility of Cooperation*. Cambridge University Press, Cambridge, 1987.

[18]   David D. Friedman. *Price Theory*. South-Western Publishing Co., Ohio, 1990.

[19]   Pamela E. Oliver and Gerald Marwell. The paradox of group size in collective action: A theory of the critical mass. ii. *American Sociological Review*, 53:1–8, 1988.

[20]   N. G. van Kampen. *Stochastic Processes in Physics and Chemistry*. North-Holland, Amsterdam, 1981.

[21]   Russell Hardin. Collective action as an agreeable n-prisoners' dilemma. *Behavioral Science*, 16(5):472–481, 1971.

[22]   H. A. Ceccatto and B. A. Huberman. Persistence of nonoptimal strategies. *Proc. Natl. Acad. Sci. USA*, 86:3443–3446, 1989.

[23]   Natalie S. Glance and Bernardo A. Huberman. Social dilemmas and fluid organizations. In K. Carley and M. Prietula, editors, *Computational Organization Theory*, pages 217–239. Lawrence Elbraum Associates, Inc., New Jersey, 1994.

[24]   Brian Dumaine. The bureaucracy busters. *Fortune*, pages 36–50, 1991.

Multi-agent planning

Emergent Constraint Satisfaction through Multi-Agent Coordinated Interaction

JyiShane Liu and Katia Sycara

The Robotics Institute
School of Computer Science
Carnegie Mellon University
Pittsburgh, PA 15213, U.S.A.
jsl@cs.cmu.edu and katia@cs.cmu.edu

Abstract. We present a methodology, called Constraint Partition and Coordinated Reaction (CP&CR), for distributed constraint satisfaction based on partitioning the set of constraints into subsets of different constraint types. Associated with each constraint type is a set of specialised agents, each of which is responsible for enforcing constraints of the specified type for the set of variables under its jurisdiction. Variable instantiation is the joint responsibility of a set of agents, each of which has a different perspective on the instantiation according to a particular constraint type and can revise the instantiation in response to violations of the specific constraint type. The final solution *emerges* through incremental local revisions of an initial, possibly inconsistent, instantiation of all variables. Solution revision is the result of coordinated local reaction of the specialised constraint agents. We have applied the methodology to job shop scheduling, an NP-complete constraint satisfaction problem. Utility of different types of coordination information in CP&CR was investigated. In addition, experimental results on a benchmark suite of problems show that CP&CR performed considerably well as compared to other centralised search scheduling techniques, in both computational cost and number of problems solved.

1   Introduction

Distributed AI (DAI) has primarily focused on Cooperative Distributed Problem Solving [4] [8] by sophisticated agents that work together to solve problems that are beyond their individual capability. Another trend has been the study of computational models of agent societies [9], composed of simple agents that interact asynchronously. With few exceptions (e.g.[1] [5] [18]), these models have been used to investigate the evolutionary behaviour of biological systems [10] [12] rather than the utility of these models in problem solving. We have developed a computational framework for problem solving by a society of simple interacting agents and applied it to solve job shop scheduling Constraint Satisfaction Problems (CSPs). Experimental results, presented in Sect. 3, show that the approach performs considerably well as compared to centralised search methods for a set of benchmark job shop scheduling problems. These encouraging results indicate good problem solving potential of approaches based on distributed agent interactions.

Many problems of theoretical and practical interest (e.g., parametric design, resource allocation, scheduling) can be formulated as CSPs. A CSP is defined by a set of *variables* $X = \{x_1, x_2, \cdots, x_m\}$, each having a corresponding domain of *values* $V = \{v_1, v_2, \cdots, v_m\}$, and a set of *constraints* $C = \{c_1, c_2, \cdots, c_n\}$. A constraint c_i is a subset of the Cartesian product $v_l \times \cdots \times v_q$ which specifies which values of the variables are compatible with each other. A solution to a CSP is an assignment of values (an instantiation) for all variables, such that all constraints are satisfied. In general, CSPs are solved by two complementary approaches, backtracking and network consistency algorithms [11][2][16]. Recently, heuristic revision [13] and decomposition [3][6] techniques for CSPs have been proposed.

Recent work in DAI has considered the distributed constraint satisfaction problem (DCSP) [22] [20] in which variables of a CSP are distributed among agents. Each agent has a subset of the variables and tries to instantiate their values. Constraints may exist between variables of different agents and the instantiations of the variables must satisfy these inter-agent constraints. In these approaches, each agent was responsible for checking that all constraints involving the values of variables under its jurisdiction were satisfied, or identifying and resolving any constraint conflicts through asynchronous backtracking. Variables were instantiated in some order, according to a static ([22]) or dynamic ([20]) variable and value ordering, and the final solution was generated by merging partial instantiations that satisfied the problem constraints. Instead, our approach decomposes a CSP by constraint type. This results in no inter-agent constraints, but each variable may be instantiated by more than one agent. While satisfying its own constraints, each agent instantiates/modifies variable values based on coordination information supplied by others. Coordination among agents facilitates effective problem solving.

In this paper, we present an approach, called Constraint Partition and Coordinated Reaction (CP&CR), in which the set of agents is partitioned into agent subsets according to the types of constraints present in the DCSP. The fundamental characteristics of CP&CR are: (1) divide-and-conquer with effective coordination (2) avoid sophisticated inter-agent interactions and rely on collective simple local reactions. CP&CR divides a Constraint Satisfaction Problem into several subproblems, each of which concerns the satisfaction of constraints of a particular type. Enforcement of constraints on variables within a subproblem is assigned to a dedicated local problem solving agent which revises variable instantiations so that its own constraints are satisfied. Since each variable may be restricted by more than one constraint type, this means that the instantiation of a variable may be changed by different local problem solving agents. Each agent is iteratively activated and examines local views of a current solution. If it does not find any conflicts in the current iteration, it leaves the current solution unchanged and terminates its own activation. If it does find local constraint violations, it changes the instantiation of one or more variables. A final solution is an instantiation of all variables that all agents agree on, i.e. it does not violate any constraints.

The remainder of the paper is organised as follows. In Sect. 2, we describe an application of CP&CR to job shop scheduling with non-relaxable time windows. In Sect. 3, we evaluate experimental results on previously studied test problems. Finally, in Sect. 4, we conclude the paper and outline our current work on CP&CR.

2 Distributed Job Shop Scheduling by CP&CR

Job shop scheduling with non-relaxable time windows involves synchronisation of the completion of a number of jobs on a limited set of resources (machines). Each job is composed of a sequence of activities (operations), each of which has a specified processing time and requires the exclusive use of a designated resource for the duration of its processing (i.e. resources have only unit processing capacity). Each job must be completed within an interval (a time window) specified by its release and due time. A solution of the problem is a schedule, which assigns start times to each activity, that satisfies all *temporal activity precedence, release and due date*, and *resource capacity* constraints. This problem is known to be NP-complete [7], and has been considered as one of the most difficult CSPs. Traditional constraint satisfaction algorithms are shown to be insufficient for this problem [17].

CP&CR views each activity as a *variable* with a *value* corresponding to the start time of the activity. Dominant constraints in job shop scheduling are partitioned into two categories: temporal precedence[1] and resource capacity constraints. Within each constraint type, subproblems are formulated. Each subproblem is assigned to a separate agent. In particular, enforcing temporal precedence constraints within an job is a subproblem that is assigned to an *job agent*; enforcing capacity constraints on a given resource is a subproblem that is under the responsibility of a *resource agent*. Therefore, for a given scheduling problem, the number of subproblems (and the number of agents) is equal to the sum of the number of jobs plus the number of resources. An activity is governed both by a job agent and a resource agent. Manipulation of activities by job agents may result in constraint violations for resource agents and vice-versa. Therefore, coordination between agents is crucial for prompt convergence on a final solution.

2.1 A Society of Reactive Agents

In CP&CR, problem solving of a job shop scheduling CSP is transformed into collective behaviours of reactive agents. Each agent examines and makes changes to only local activities under its responsibility, and seeks for satisfaction by ensuring that no constraint in its assigned constraint subset is violated. Agents

---

[1] Release and due dates constraints are considered as temporal precedence constraints between activities and fixed time points.

are equipped with only primitive behaviour. When activated, each agent goes through an Examine-Resolve-Encode cycle (see Fig. 1). It first examines its local view of current solution, i.e. the values of the variables under its jurisdiction. If there are constraint violations, it changes variable instantiations to resolve conflicts according to innate heuristics and coordination information (see Sect. 2.2 and 2.3).

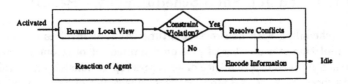

Fig. 1. Agent's reactive behaviour

Agents coordinate by *passive* communication. They do not communicate with each other directly. Instead, each agent reads and writes coordination information on variables under its jurisdiction. Coordination information on a variable represents an agent's partial "view" on the current solution and is consulted when other agents are considering changing the current instantiation of the variable to resolve their conflicts. Each time an agent is activated and has ensured its satisfaction, it writes down its view on current instantiations on each variable under its jurisdiction as coordination information.

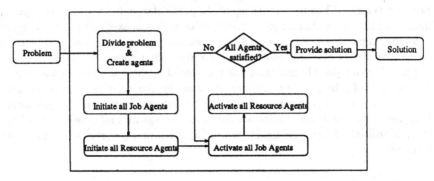

Fig. 2. System Control Flow

System initialisation is done as follows: (1) decomposition of the input scheduling problem according to resource and job constraints, (2) creation of the corresponding resource and job agents, (3) activation of the agents (see Fig. 2). Initially each job agent calculates boundary[2] for each activity under its ju-

---

[2] The boundary of an activity is defined as the interval between its earliest possible start time and its latest possible finish time.

risdiction considering its release and due date constraints. Each resource agent calculates the contention ratio for its resource by summing the durations of activities on the resource and dividing by the interval length between the earliest and latest time boundary among the activities. If this ratio is larger than a certain threshold, a resource agent concludes that it is a bottleneck resource agent.[3][4]

Activities under the jurisdiction of a bottleneck resource agent are marked as *bottleneck activities* by the agent. Each resource agent heuristically allocates the earliest free resource interval to each activity under its jurisdiction according to each activity's boundary. After the initial activation of resource agents, all activities are instantiated with a start time. This initial instantiation of all variables represents the initial configuration of the solution.[5]

Subsequently, job agents and resource agents engage in an evolving process of reacting to constraint violations and making changes to the current instantiation. In each *iteration cycle*, job and resource agents are activated alternatively, while agents of the same type are activated simultaneously, each working independently. When an agent finds constraint violations under its jurisdiction, it employs local reaction heuristics to resolve the violations. The process stops when none of the agents detect constraint violations during an iteration cycle. The system outputs the current instantiation of variables as a solution to the problem.

2.2 Coordination Information

Coordination information *written* by a job agent on an activity is referenced by a resource agent, and vice-versa.

Job agents provide the following coordination information for resource agents.

1. *Boundary* is the interval between the earliest start time and latest finish time of an activity (see Fig. 3). It represents the overall temporal flexibility of an activity and is calculated only once during initial activation of job agents.
2. *Temporal Slack* is an interval between the current finish time of the previous activity and current start time of the next activity (see Fig. 3). It indicates the temporal range within which an activity may be assigned to without causing temporal constraint violations. (This is not guaranteed since temporal slacks of adjacent activities are overlapping with each other.)

---

[3] If no bottleneck resource is identified, threshold value is lowered until the most contended resource is identified.

[4] In job shop scheduling, the notion of bottleneck corresponds to a particular resource interval demanded by activities that exceeds the resource's capacity. Most state-of-the-art techniques emphasise the capability to identify *dynamic* bottlenecks that arise during the construction of solution. In our approach, the notion of bottleneck is *static* and we exploit the dynamic local interactions of agents.

[5] We have conducted experiments with random initial configurations and confirmed that CP&CR is barely affected by its starting point, i.e. CP&CR has equal overall performance with heuristic and random initial configurations.

112

3. *Weight* is the weighted sum of relative temporal slack with respect to activity boundary and relative temporal slack with respect to the interval bound by the closest bottleneck activities (see Fig. 4). It is a measure of the likelihood of the activity "bumping" into an adjacent activity, if its start time is changed. Therefore, a high weight represents a job agent's preference for not changing the current start time of the activity. In Fig. 4, activity-p of job B will have a higher weight than that of activity-a of job A. If both activities use the same resource and are involved in a resource capacity conflict, the resource agent will change the start time of activity-a rather than start time of activity-p.

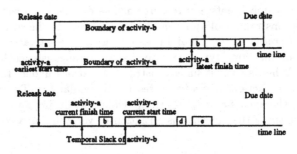

Fig. 3. Coordination information: Boundary and Temporal Slack

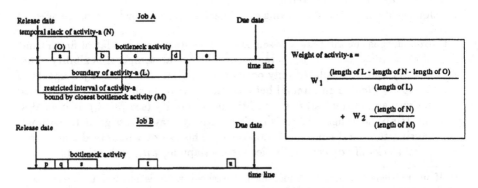

Fig. 4. Coordination information: Weight

Resource agents provide the following coordination information for job agents.

1. *Bottleneck Tag* is a tag which marks that this activity uses a bottleneck resource. This tag is put by a bottleneck resource agent on all activities

under its jurisdiction. It implies that job agent should treat these activities differently.

2. *Resource Slack* is an interval between the current finish time of the previous activity and the current start time of the next activity on the resource timeline (see Fig. 5). It indicates the range of activity start times in which an activity may be changed without causing capacity constraint violations. (There is no guaranteed since resource slacks of adjacent activities are overlapping with each other.)

3. *Change Frequency* is a counter of how frequently the start time of this regular activity set by a job agent is changed by a submissive resource agent. It measures the search effort of job and regular resource agents to evolve an instantiation on regular activities that is compatible to the current instantiation on bottleneck activities.

Resource X

time line

resource slack of activity-a

Fig. 5. Coordination information: Resource Slack

2.3 Reaction Heuristics

Agents' reaction heuristics utilise perceived coordination information and incorporate coordination strategies of group behaviours. We have developed coordination strategies that promote rapid convergence by minimising the ripple effects of causing conflicts to other agents as a result of an agent's fixing the current constraint violations. Conflict minimisation is achieved by minimising the number and extent of activity start time changes. Other coordination strategies include using most constrained agents as an anchor of interaction, and preventing oscillatory value changes.

Reaction Heuristics of Job Agent. Job agents resolve conflicts by considering conflict pairs. A conflict pair involves two adjacent activities whose current start times violate the precedence constraint between them (see Fig. 6). Conflict pairs are resolved one by one. A conflict pair involving a bottleneck activity, i.e., an activity with tighter constraints, is given a higher conflict resolution priority. To resolve a conflict pair, job agents essentially determine which activity's current start time should be changed. If a conflict pair includes a bottleneck and a regular activity, depending on whether the change frequency counter on the regular activity in the conflict pair is still under a threshold, job agents change the start time of either the regular or the bottleneck activity. For conflict pairs of regular activities, job agents take into consideration additional factors, such as value changes feasibility of each activity, change frequency, and resource slack.

In Fig. 6, the conflict pair of activity-A2 and activity-A3 will be resolved first since activity-A2 is a bottleneck activity. If the change frequency of activity-A3 is still below a threshold, start time of activity-A3 will be changed by an addition of T2 (the distance between current start time of activity-A3 and current end time of activity-A2) to its current start time. Otherwise, start time of activity-A2 will be changed by a subtraction of T2 from its current start time. In both cases, start time of activity-A4 will be changed to the end time of activity-A3. To resolve the conflict pair of activity-A0 and activity-A1, either start time of activity-A0 will be changed by a subtraction of T1 from its current start time or start time of activity-A1 will be changed by an addition of T1 to its current start time. If one of the two activities can be changed within its boundary and resource slack, job agent A will change that activity. Otherwise, job agent A will change the activity with less change frequency.

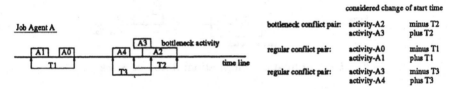

Fig. 6. Conflict Resolution of Job Agent

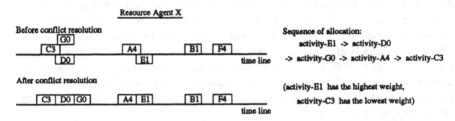

Fig. 7. Conflict Resolution of Regular Resource Agent

Reaction Heuristics of Regular Resource Agents. To resolve constraint violations, resource agents re-allocate the over-contended resource intervals to the competing activities in such a way as to resolve the conflicts and, at the same time, keep changes to the start times of these activities to a minimum. Activities are allocated according to their weights, boundaries, and temporal slacks. Since an activity's weight is a measure of the desire of the corresponding job agent to keep the activity at its current value, activity start time decisions based on weight reflect group coordination. For example, in Fig. 7, activity-A4 was preempted by activity-E1 which has higher weight. Start time of activity-A4 is changed as little as possible. In addition, when a resource agent perceives a high resource contention during a particular time interval (such as the conflict involving activity-C3, activity-D0, and activity-G0), it allocates the resource intervals and assigns high change frequency to these activities, and thus dynamically changes the priority of these instantiation.

Reaction Heuristics of Bottleneck Resource Agents. A bottleneck resource agent has high resource contention. This means that most of the time a bottleneck resource agent does not have resource slack between activities. When the start time of a bottleneck activity is changed, capacity constraint violations are very likely to occur. A bottleneck resource agent considers the amount of overlap of activity resource intervals on the resource to decide whether to right-shift some activities (Fig. 8 (i)) or re-sequence some activities according to their current start times by swapping the changed activity with an appropriate activity. (Fig. 8 (ii)). The intuition behind the heuristics is to keep the changes as minimum as possible.

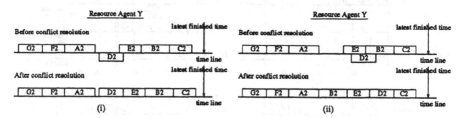

Fig. 8. Conflict Resolution of Bottleneck Resource Agent

2.4 Solution Evolution

Figure 9 shows a solution evolution process of a very simple problem where resource Y is regarded as a bottleneck resource. In (a), resource agents allocate their earliest possible free resource intervals to activities, and thus construct the initial configuration of variable instantiation. In (b), A13, A23 within dotted rectangular boxes represent the start times assigned by resource agents Res.X and Res.Z, respectively. Job1 and Job2 agents are not satisfied with current instantiation because the pairs of (A12 A13) and (A22 A23) are violating their precedence constraints. Job1 (cf. Job2) agent changes the start times of A13 (cf. A23) (shown by solid rectangular box) because A12 (cf. A22) is a seed activity and change frequency of A13 (cf. A23) is zero (have not exceed the threshold). In (c), Res.Z agent finds a capacity constraint violation between A23 and A33 (shown by dotted rectangular box before conflict resolution), and changes the start time of A33 because A23 has a higher weight. All agents are satisfied with the current instantiation of variables in (d) which represents a solution to the problem.

Figure 10 shows a solution evolution process in terms of occurred conflicts for a more difficult problem which involves 10 jobs on 5 resources. In cycle 0, resource agents construct an initial instantiation of variables that includes start times of bottleneck activities by bottleneck resource agents. During cycle 1 to cycle 9, job agents and regular resource agents try to evolve a compatible instantiation of regular activities with the instantiation of bottleneck activities. In cycle 10, some job agents perceive the effort as having failed (by observing the change frequency counter on a regular variable in conflict with a bottleneck activity has exceeded the threshold) and change the values of their bottleneck activities. Bottleneck resource agents respond to constraint violations by modifying instantiation of the bottleneck activities. This results in a sharp increase of conflicting activities for job agents in cycle 11. Again, the search for compatible instantiation resumes until another modification on instantiation of the bottleneck activities in cycle 16. In cycle 18, the solution is found.

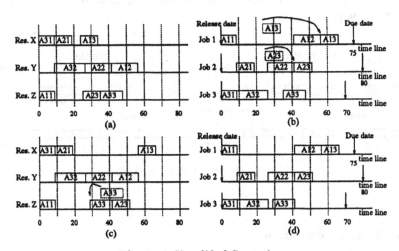

Fig. 9. A Simplified Scenario

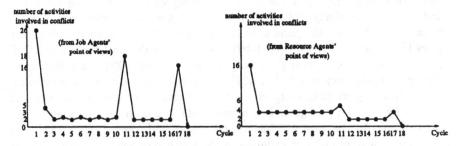

Fig. 10. Conflicts Evolution of a more difficult problem

3 Evaluation on Experimental Results

We evaluated the performance of CP&CR on a suite of job shop scheduling CSPs proposed in [17]. The benchmark consists of 6 groups, representing different scheduling conditions, of 10 problems, each of which has 10 jobs of 5 activities and 5 resources. Each problem has at least one feasible solution. CP&CR has been implemented in a system, called CORA (COordinated Reactive Agents). We experimentally (1) investigated the effects of coordination information in the system, (2) compared CORA's performance to other constraint-based as well as priority dispatch scheduling methods.

3.1 Effects of Coordination Information

In order to investigate the effects of coordination information on the system's performance, we constructed a set of four coordination configurations.

- C0 represents a configuration in which the system ran with no coordination information at all. Without boundary information, when initially activated, resource agents allocate resource intervals according to random sequences. When job agents are activated, they resolve conflicts by randomly changing the instantiation of one of the two activities in each conflict pair. Similarly, resource agents resolve conflicts based on random priority sequences.
- C1 represents a configuration in which only boundary information is available. Resource agents use this information for heuristic initial allocation of resource intervals. After the initial schedule is generated, no other information is available for conflict resolutions.
- C2 represents a configuration in which boundary and bottleneck tag information is available. Resource agents use the boundary information for heuristic initial allocation of resource intervals. Job agents use the bottleneck tag information to bias resolution of conflict pairs.
- C3 represents a complete configuration in which all coordination information is provided for resource agents and job agents.

Figure 11 shows the comparative performance of different configurations on the suite of benchmark problems. The additional coordination information for each configuration is underlined in Fig. 11 (i). The number of cycles that the system was allowed was limited to 100. If there were still conflicts at cycle 100, the system gave up solving the problem. Since system operations in C0, C1, and C2 have random nature, they were ran on each problem 10 times. The numbers reported are the average number, e.g. 15.8 out of 60 problems were solved means that there were 158 successful runs among 600 (10 runs for each problem). C3 is deterministic and for it each problem was tried only once. We confirm that adding coordination information enables the system to solve more problems within fewer cycles. The results shows the utility of coordination information.

Figure 11 (ii) shows, for different coordination configurations, the successful overall problem solving processes[6] in terms of the number of activities involved in conflicts at each cycle. As the coordination information increases, the shape of the curve indicates a steeper drop in the number of conflicts in fewer cycles. This indicates that increasing rates of convergence are facilitated by more coordination information. The curve for deterministic C3 has a peak at cycle 5. This reveals that when the problem was not solved within the first few cycles, an instantiation modification on the activities using bottleneck resources typically occurred. The curves for C0, C1, and C2 do not exhibit a peak because the system does not have particular pattern of interaction in those coordination configurations.

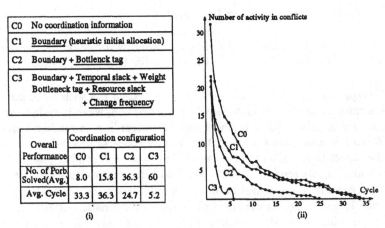

C0	No coordination information
C1	Boundary (heuristic initial allocation)
C2	Boundary + Bottlenck tag
C3	Boundary + Temporal slack + Weight Bottleneck tag + Resource slack + Change frequency

Overall Performance	Coordination configuration			
	C0	C1	C2	C3
No. of Porb Solved(Avg.)	8.0	15.8	36.3	60
Avg. Cycle	33.3	36.3	24.7	5.2

(i)

(ii)

Fig. 11. Comparative Performance between Coordination Configurations

3.2 Comparison with Other Scheduling Techniques

CORA was compared to four other heuristic search scheduling techniques, ORR /FSS, MCIR, CPS, and PCP. ORR/FSS [17] incrementally constructs a solution by chronological backtracking search guided by specialised variable and value ordering heuristics. ORR/FSS+ is an improved version augmented with an intelligent backtracking technique [21]. Min-Conflict Iterative Repair (MCIR) [13] starts with an initial, inconsistent solution and searches through the space of possible repairs based on a *min-conflicts* heuristic which attempts to minimise the number of constraint violations after each step. Conflict Partition Scheduling (CPS) [15] employs a search space analysis methodology based on stochastic simulation which iteratively prunes the search space by posting additional

---

[6] For C0, C1, and C2, only successful overall problem solving processes are averaged and shown.

constraints. Precedence Constraint Posting (PCP) [19] conducts the search by establishing sequencing constraints between pairs of activities using the same resource based on *slack-based* heuristics. In addition, three frequently used and appreciated priority dispatch rules from the field of Operations Research: EDD, COVERT, and R&M [14], are also included for comparison.

Table 1 reports the number of problems solved[7] and the average CPU time spent over all the benchmark problems for each technique. Note that the results of ORR/FSS, ORR/FSS+, MCIR, CPS, and PCP were obtained from published reports, of mostly the developers of the techniques. MCIR is the only exception, which is implemented by Muscettola who reported its results based on randomly generated initial solutions [15]. All CPU times were obtained from Lisp implementations on a DEC 5000/200. In particular, CORA was implemented in CLOS (Common Lisp Object System). CPS, MCIR, ORR/FSS, and ORR/FSS+ were implemented using CRL (Carnegie Representation Language) as an underlying frame-based knowledge representation language. CPU times of CPS, MCIR, ORR/FSS, and ORR/FSS+ were divided by six from the published numbers as an estimate of translating to straight Common Lisp implementation.[8] PCP's CPU times are not listed for comparison because its CPU times in Lisp are not available. Its reported CPU times in C are 0.3 second [19]. Although CORA can operate asynchronously, it was sequentially implemented for fair comparison. The results show that CORA works considerably well as compared to the other techniques both on feasibility and efficiency in finding a solution.

	CORA	CPS	MCIR	ORR/ FSS	ORR/ FSS+	PCP	EDD	COVERT	R&M
w/1	10	10	9.8	10	10	10	10	8	10
w/2	10	10	2.2	10	10	10	10	7	10
n/1	10	10	7.4	8	10	10	8	7	9
n/2	10	10	1	9	10	10	8	6	9
o/1	10	10	4.2	7	10	10	3	4	6
o/2	10	10	0	8	10	8 ~ 10	8	8	8
Total	60	60	24.6	52	60	58 ~ 60	47	40	52
AVG. CPU time	4.8 seconds	13.07 * seconds	49.74 * seconds	39.12 * seconds	21.46 * seconds	N/A	0.9 seconds	0.9 seconds	0.9 seconds

Table 1. Performance Comparison

3.3 Evaluation

As a scheduling technique, CORA performs a heuristic *approximate* search in the sense that it does not systematically try all possible configurations. Al-

---

[7] PCP's performance is sensitive to the parameters that specify search bias [19].

[8] ORR/FSS and ORR/FSS+ obtained 30 times speedup in C/C++ implementation. We assumed a factor of five between Common Lisp and C/C++ implementations.

though there are other centralised scheduling techniques that employ similar search strategies, CORA distinguishes itself by an interaction driven search mechanism based on well-coordinated asynchronous local reactions. Heuristic approximate search provides a middle ground between the generality of domain-independent search mechanisms and the efficiency of domain-specific heuristic rules. Instead of the rigidity of one-pass attempt in solution construction (either it succeeds or fails, and the decisions are never revised) in approaches using heuristic rules, CORA adapts to constraint violations and performs an effective search for a solution. As opposed to generic search approaches, in which a single search is performed on the whole search space and search knowledge is obtained by analysing the whole space at each step, CORA exploits local interactions by analysing problem characteristics and conducts well-coordinated asynchronous local searches.

The experimental results obtained by various approaches concur with the above observations. Approaches using generic search techniques augmented by domain-specific search-focus heuristics (ORR/FSS, ORR/FSS+, MCIR, CPS) required substantial amount of computational effort. Some of them could not solve all problems in the sense that they failed to find a solution for a problem within the time limit set by their investigators. Approaches using dispatch rules (EDD, COVERT, R&M) were computationally efficient, but did not succeed in all problems. PCP relies on heuristic rules to conduct one-pass search and its performance is sensitive to parameters that specify search bias. CORA struck a good balance in terms of solving all problems with considerable efficiency. Furthermore, with a mechanism based on collective operations, CORA can be readily implemented in parallel processing such that only two kinds of agents are activated sequentially in each iteration cycle, instead of 10 job agents and 5 resource agents under current implementation. This would result in an approximate time-reducing factor of 7 (i.e., 15/2) and would enable CORA to outperform all other scheduling techniques in comparison.

4   Conclusions

We have presented an approach to distributed constraint satisfaction based on partitioning the problem constraints into constraint types. Responsibility for enforcing constraints of a particular type is given to specialist agents. The agents coordinate to iteratively change the instantiation of variables under their jurisdiction according to their specialised perspective. The final solution *emerges* through incremental local revisions of an initial, possibly inconsistent, instantiation of all variables. We demonstrated the effectiveness of the approach in the domain of job shop scheduling. The power of our approach stems from the coordinated local interactions of the reactive agents. We are currently formalising the CP&CR methodology and extending it to Constraint Optimisation Problems (COPs). We are also investigating the utility of CP&CR in other domains with different problem structures.

References

1. Brooks, R.: Intelligence Without Reason. In Proceedings of the IJCAI-91. (1991) 569–595
2. Dechter, R.: Network-based heuristics for constraint satisfaction problems. Artificial Intelligence. **34** (1988) 1–38
3. Dechter, R., Meiri, I., Pearl, J.: Temporal constraint networks. Artificial Intelligence. **49** (1991) 61–95
4. Decker, K.: Distributed problem-solving techniques: A survey. IEEE Transactions on Systems, Man, and Cybernetics. **17** (1987) 729–739
5. Ferber, J., Jacopin, E.: The framework of Eco problem solving. In Demazeau and Muller, editors, Decentralized AI II. (1991) Elsevier, North-Holland.
6. Freuder, E., Hubbe, P.: Using inferred disjunctive constraints to decompose constraint satisfaction problems. In Proceedings of the IJCAI-93. (1993) 254–260
7. Garey, M., Johnson, D.: Computers and Intractability: A Guide to the Theory of NP-Completeness. (1979) Freeman and Co.
8. Gasser, L., Hill, R., Jr.: Engineering coordinated problem solvers. Annual Review of Computer Science. **4** (1990) 203–253
9. Langton, C., editor: Artificial Life. (1989) Addison-Wesley.
10. Langton, C., Taylor, C., Farmer, J., Rasmussen, S., editors: Artificial Life II. (1991) Addison-Wesley.
11. Mackworth, A.: Constraint satisfaction. In Shapiro, S., editor, Encyclopedia in Artificial Intelligence. (1987) 205–211. Wiley, New York.
12. Meyer, J.-A., Wilson, S., editors: Proceedings of the First International Conference on Simulation of Adaptive Behaviour - From Animals To Animats. (1991) MIT Press.
13. Minton, S., Johnston, M., Philips, A., Laird, P.: Minimising conflicts: a heuristic repair method for constraint satisfaction and scheduling problems. Artificial Intelligence. **58** (1992) 161–205
14. Morton, T., Pentico, D.: Heuristic Scheduling Systems: With Applications to Production Systems and Project Management. (1993) John Wiley & Sons, New York.
15. Muscettola, N.: HSTS: Integrated planning and scheduling. In Fox, M., Zweben, M., editors. Knowledge-Based Scheduling. (1993) Morgan Kaufmann.
16. Nadel, B.: Constraint satisfaction algorithms. Computational Intelligence. **5** (1989) 188–224
17. Sadeh, N.: Look-ahead techniques for micro-opportunistic job shop scheduling. Technical report CMU-CS-91-102, School of Computer Science, Carnegie-Mellon University. (1991)
18. Shoham, Y., Tennenholtz, M.: On the synthesis of useful social laws for artificial agent societies. In Proceedings of AAAI-92. (1992) 276–281
19. Smith, S., Cheng, C.: Slack-based heuristics for constraint satisfaction scheduling. In Proceedings of AAAI-93. (1993) 139–144
20. Sycara, K., Roth, S., Sadeh, N., Fox, M.: Distributed constraint heuristic search. IEEE Transactions on System, Man, and Cybernetics. **21** (1991) 1446–1461
21. Xiong, Y., Sadeh, N., Sycara, K.: Intelligent backtracking techniques for job shop scheduling. In Proceedings of the Third International Conference on Principles of Knowledge Representation and Reasoning. (1992) 14–23
22. Yokoo, M., Ishida, T., Kuwabara, K.: Distributed constraint satisfaction for DAI problems. In Proceedings of the 10th International Workshop on Distributed AI. (1990)

Sophisticated and Distributed:
The Transportation Domain

– Exploring Emergent Functionality in a Real-World Application –

K. Fischer, N. Kuhn, H. J. Müller, J. P. Müller, M. Pischel

DFKI, Stuhlsatzenhausweg 3, D-66123 Saarbrücken

Abstract. In this paper, we present the MARS multi-agent system. MARS models a society of cooperating transportation companies. Emphasis is placed on how the functionality of the system as a whole - the solution of the global scheduling problem - emerges from local decision-making and problem-solving strategies, and on how variations of these strategies influence the performance of the system. We address three techniques of Distributed Artificial Intelligence (DAI) which are used for tackling the hard problems that occur in this domain, and which together give rise to the emergence of a solution to the global scheduling problem: (1) cooperation among the agents, (2) task decomposition and task allocation, and (3) decentralised planning. Finally, we briefly describe the implementation of the system and provide experimental results which show how different strategies for task decomposition and cooperation influence the behaviour of the system.

1   Introduction

Today, Distributed Artificial Intelligence (DAI) is rightfully regarded as one of the most dynamic branches within AI research. Using DAI techniques such as cooperation [6, 14], negotiation [27, 25, 17], task decomposition, and task allocation [8, 16], is believed to be promising for solving problems which are computationally and structurally complex, highly dynamic, which are characterised by incomplete and inconsistent knowledge, and where information and control are physically distributed. However, verifying this belief by implementing real-world applications remains a challenge for researchers in DAI [22].

In this paper, we explore the usefulness of several DAI techniques for modelling a real-world application domain: a scenario of transportation companies is described. The companies have to carry out transportation orders which arrive dynamically. For this purpose, they have a set of trucks at their disposal. We evaluate the behaviour of the system as a whole in a straightforward manner: the measure of coherence is how well it can solve the problem of scheduling the orders, i.e. what cost are caused by carrying out the orders. What is extraordinary with our approach is that the companies themselves do not have facilities for planning orders. It is only the trucks which maintain local plans. The actual solution to the global order scheduling problem emerges from the local decision-making of the agents. There are three specific techniques used in order to bring about this emergent functionality:

- Task decomposition and task allocation is done in order to assign orders to appropriate trucks. Different models of task decomposition and task allocation are discussed in section 3.3
- Cooperation (1) among shipping companies and (2) between a company and its trucks helps solving the problem of task decomposition and allocation. Whereas the latter kind of cooperation can be implemented using a simple contract-net-like protocol [8], cooperation among shipping companies has to respect the autonomy of the single companies, and thus requires a full model of negotiation [3, 16].
- Task decomposition and task allocation is guided by local decision criteria. These criteria are derived from the local plans. Moreover, polynomial algorithms are used in order to solve the scheduling problem locally (see subsection 3.1).

In this paper, we do not treat in a detailed manner questions of protocols for cooperation and negotiation (see [16]). Rather, we provide an empirical investigation of how different forms of cooperation and different models of task decomposition lead to different solutions to the global scheduling problem.

The paper is organised as follows: In section 2.1, we present the transportation domain. We show the relevance of the domain, and we argue for choosing a multi-agent approach for modelling the domain. In section 3, three important DAI techniques are addressed: cooperation, task decomposition and allocation, and decentralised planning. The cooperation mechanisms are explained in more detail by means of examples in section 4. Finally, in section 5, a series of experiments is described, and results are are provided and discussed.

2 The MARS Multi-agent Scenario

2.1 The Domain of Application

In a time of constantly growing world-wide economical transparency and interdependency, logistics and the planning of freight transports get more and more important both for economical and ecological reasons. Many of the problems which must be solved in this area, such as the Travelling Salesman and related scheduling problems, are known to be $\mathcal{NP}$-hard. Moreover, not only since *just-in-time* production has come up, planning must be performed under a high degree of uncertainty and incompleteness, and it is highly dynamic. Standard Operations Research approaches (see [18, 7, 26] for an overview) can hardly cope with the dynamics of this domain (see [10] for a discussion of more recent approaches to fleet scheduling). In fact, also in reality these problems are far from being solved. Recent analysis [23] has revealed that more than one out of three trucks in the streets of Europe is driving without carriage, since it is on its way to pick up goods or on its way back home.

2.2 The (D)AI Aspects

Why is it adequate to use AI techniques and more specifically DAI approaches to tackle the transportation problems described above? One reason is the complexity of the scheduling problem, which makes it very attractive for AI research[1]. However there are more pragmatic reasons: *Commonsense knowledge* (e.g. taxonomical, topological, temporal, or expert knowledge) is necessary to solve the scheduling problems effectively. *Local knowledge about the capabilities* of the transportation company as well as knowledge about competitive (and maybe cooperative) companies massively influences the solutions. Moreover, since a global view is impossible (because of the complexity), there is a need to operate from a local point of view and thus to deal with *incomplete knowledge* with all its consequences.

The last aspect leads to the DAI arguments:

1. The domain is inherently distributed. Hence it is very natural to look at it as a multi-agent system. However, instead of tackling the problem from the point of view of the entities which are to be modelled and then relying on the emergence of the global solution, the classical approach to the problem is an (artificially) centralised one.
2. The task of a centrally maintaining and processing the knowledge about the shipping companies, their vehicles, and behaviour is very complex. Moreover, knowledge is often not even centrally available (real-life company are not willing to share *all* their local information with other companies). Therefore, modelling the companies as independent and autonomous units seems the only acceptable way to proceed.
3. In real business, companies usually solve capacity problems by contacting partners that might be able to perform the problematic tasks. Then the parties negotiate the contract. However, task allocation, contracting, negotiating and performing joint actions are main topics in DAI research.

2.3 The Scenario

The MARS scenario (Modelling a Multi-Agent Scenario for Shipping Companies) [2] implements a group of shipping companies whose goal it is to deliver a set of dynamically given orders, satisfying a set of given time and/or cost constraints[2]. The complexity of the orders may exceed the capacities of a single company. Therefore, cooperation between companies is required in order to achieve the goal in a satisfactory way. The common use of shared resources, e.g. train or ship, requires coordination between the companies. Although each company has a local, primarily self-interested view, cooperation between the shipping companies is necessary in order to achieve reasonable global plans (see section 5).

---

[1] At this year's International Conference on AI and Applications (CAIA'93), seven out of sixty-one papers dealt with scheduling problems!

[2] MARS has been implemented for UNIX using the rule-based development tool MAGSY [11].

Apart from internal *system agents*, which perform tasks such as the representation and visualisation of the simulation world, the MARS agent society consists of two sorts of *domain agents*, which correspond to the logical entities in the domain: *shipping companies* and *trucks*. Looking upon trucks as agents allows us to delegate problem-solving skills to them (such as route-planning and local plan optimisation). Communication between agents is enabled by direct communication channels.

The *company* agent is responsible for the disposition of the orders that have been confided to him. Thus, it has to allocate the orders to its trucks, while trying to satisfy the constraints provided by the user as well as local optimality criteria. The shipping companies can be regarded as experts for cooperation and cooperative problem solving. They are equipped with additional global knowledge which is needed for cooperating successfully with other companies.

The *truck* agents represent the means of transport of a transportation company. Each truck agent is associated with a particular shipping company from which it receives orders of the form *"Load a goods g_1 at location l_1 and transport it to location l_2"*. Given such an order, the truck agent does the planning of the route ([15], see also section 3.1) according to its geographical knowledge and it will inform the shipping company agent about the deliverance of the goods. Furthermore, it is able to support the shipping company during the disposition phase: The truck reports remaining capacities, planned routes and it is able to estimate the effort (and the effects)[3] that are caused by an order.

3 DAI Techniques in the MARS Scenario

The description of the scenario reveals the autonomy of the agents as a necessary condition for a modelling that reflects the real world situation and that can even support the dispatcher in a real shipping company. In this section we describe several DAI methods that are used within MARS scenario.

The general idea of the solution is based on the paradigm of self-organisation, which is applied to a society of knowledge-based systems: at the beginning, the system is in a state of equilibrium. This equilibrium is disturbed by the distribution of a set of orders among the company agents. This stimulates the truck agents to devise local plans and to inform their company about the cost arising for carrying out an order in their local context. Based on this information the company agent allocates the orders to its trucks. Having done this, the society of agents has constructed a valid plan to deliver the set of initial orders and has reached a state of equilibrium in the sense of [24]. Following the terminology of Steels further on, there are two types of dynamics which can disturb this equilibrium: an internal dynamics which is due to that the trucks reflect on their plans, and an external dynamics caused by incoming orders. Both kinds of dynamic events result in message passing activities in the system. By passing messages among the agents, local disturbances spread out into the local plans

---

[3] i.e. cost, time, security of transport, ...

of other agents. To reach an equilibrium state again (i.e., another valid global plan for the actual set of orders) the messages are structured into *negotiation protocols*. The dissipative structure described by the different protocols is a co-operative task decomposition process based on the exchange of orders between the agents implementing a distributed and decentralised scheduling algorithm for this application domain.

For the rest of this section we stress three aspects of this approach: the *planning* of the single agents, the forms of *cooperation*, and the mechanisms for *task decomposition*.

3.1 The Distributed Scheduling Approach

The task of delivering several orders is basically a scheduling problem. What makes it even harder is the two-dimensionality of task decomposition resulting from the special domain. The goods to be transported can be distributed to several means of transport (truck, train, ship, plane), and the route between two cities on a road map can be splitted up into sub-routes which can be taken at different times using different conveyances. Due to the combinatorial explosion resulting from this, it is often impossible to devise a globally optimal plan. We tackle this problem by computing locally good solutions for each agent. Thus, we hope to get an reasonable overall solution for the given problem. This solution is further optimised by cooperation between the problem solving agents. Here two agents only agree to a solution if none of them gets a decrease in his local utility, and if at least one of them has an increase in utility by the deal. This process of negotiation leads to pareto-optimal solutions.

The problem of allocating a set of orders to a set of trucks is an $\mathcal{NP}$-hard problem. This can be shown by reducing the Rural Postman Problem (cf. [12]) which is known to be $\mathcal{NP}$-complete to the decision problem which corresponds to the order allocation problem. The Rural Postman Problem is defined as follows:

> **INSTANCE**: Graph $G = (V, E)$, length $l(e) \in Z_0^+$ for each $e \in E$, subset $E' \subseteq E$, bound $B \in Z^+$.
> **QUESTION**: Is there a circuit in G that includes each edge in E' and that has total length of at most B?

The Rural Postman Problem remains $\mathcal{NP}$-complete even if $l(e) = 1$ for all $e \in E$. The relationship between the modified problem and the task allocation problem becomes clear when we consider a company agent who only possesses a single truck, and who has for each edge $e_i = (v_{i_1}, v_{i_2}) \in E'$ an order o_i from location v_{i_1} to location v_{i_2} which needs the whole capacity of the truck. Then, there exists a circuit in G including each edge in E' which has a total length of at most B iff there exists a route for the truck which is at most of length B.

This reduction shows that both the task allocation problem and route planning of the trucks for a set of orders are $\mathcal{NP}$-hard problems (see [9] for the detailed proof). Thus, in order to keep them manageable in a computer implementation, heuristic algorithms have to be applied which do not guarantee optimality, but which in most cases provide good results in a reasonable amount of time.

The solution for the routing of the trucks is built up incrementally. If a truck has a plan to visit the locations $l_0, \ldots, l_n, l_0$ in order to deliver orders $o_1, \ldots, o_m$ and has to add a new order o from starting location s to target location t it inserts s and t into his plan such that the detour is minimised. This does not mean that s and t need not to be successive locations in the new plan. Rather, enough capacity has to remain to deliver the orders $o_{i+1}, \ldots, o_{j+1}$ together with order o. If s is inserted after l_i and t is inserted after l_j with $j > i$, the detour is computed by the formula if $j > i + 1$

$$detour = \begin{cases} dist(l_i, s) \ + dist(s, l_{i+1}) \ + dist(l_j, t) \ + dist(t, l_{j+1}) \\ \qquad - (dist(l_i, l_{i+1}) + dist(l_j, l_{j+1})) \qquad \text{if } j > i + 1 \\ \\ dist(l_i, s) \ + dist(s, t) \ + dist(t, l_{i+1}) - dist(l_i, l_{i+1}) \quad \text{if } j = i + 1 \end{cases}$$

The time needed by this algorithm to insert one order into a delivery plan containing n locations to visit is bound by $\mathcal{O}(n^2)$. The number of locations in a plan depends linearly on the number of orders a truck has got. Therefore, by this algorithm the time needed for the planning of a route for m orders is bound by $\mathcal{O}(m^3)$. The allocation of the orders to trucks by the company agent is done using the contract net protocol: the company agent offers an order to some eligible trucks who evaluate their plans and inform the company agent about that. Based on this information it chooses the best offer and allocates the order to that truck. It follows from the above considerations that this algorithm for the task allocation within a shipping company is also of time complexity $\mathcal{O}(m^3)$.

Another interesting question is what kind of algorithm is implemented by this procedure. However, this depends on the company agent's strategy for processing the m contract nets to allocate the orders: if it always completes a protocol before it initiates the one for the next order, the procedure described above implements a *greedy* algorithm for the task allocation process. An alternative to that strategy is that the company agent can maintain several contract net protocols for different orders at the same time and can base its decision on the information that it receives by all the bids in the different protocols. Thus, the strategies of the agent allow to incorporate a broad range of heuristics into the allocation process. However, the second alternative requires more sophisticated planning mechanisms, since there is high uncertainty in the plans. For instance, if an order is open for allocation and the truck is asked to give a bid for a second order, *shall it believe that it is allocated the first one or not?* The different assumptions lead to different bids. Therefore, in our system, we preferred the former strategy.

The outcome of the task allocation can be improved by the cooperation between the different companies. A truck has some knowledge to find out weak points in his plan, e.g. empty rides or orders that lead to large detours. By telling its company agents about this he can initiate initiate different cooperation mechanisms by which the plans can be improved due to an exchange of orders between the agents.

3.2 Cooperation Settings

In the previous subsection, the local algorithms used by the truck agents were described. In order to coordinate the local activities, and in order to achieve a coherent global behaviour of the system, the agents have to cooperate. In this subsection, we define three basic cooperation settings, namely *vertical cooperation*, *horizontal cooperation*, and *enhanced cooperation*, which we implemented in the MARS domain.

Vertical Cooperation (VC) *Vertical cooperation* describes the process of task decomposition and task allocation between a shipping company and its trucks. This relation is hierarchical; the trucks are obliged to give their best to carry out orders given by their company, and they are obliged to provide the company with important information upon request. We use a slight variation of the contract net [8] in order to model this kind of interaction.

More precisely the procedure is as follows: The shipping company partitions orders into cargos that can be transported by single trucks. These cargos are offered to the trucks of the company. The bid of a truck describes the costs that will arise for it when carrying out this order. According to local decision functions and the incoming bids of the trucks the company allocates the subtasks to some of the trucks. This basic solution closely corresponds to the *centralised model of task decomposition* which is described in section 3.3. An advanced release of the protocol allows the trucks to bid also for only a part of an order; this is one step towards a decentralisation of task decomposition.

Horizontal Cooperation (HC) *Horizontal cooperation* means cooperation among a group of autonomous shipping companies. Companies can exchange orders and information about free loading capacity. This exchange, however, is not performed hierarchically; rather, it reveals all aspects of conflict, competition and cooperation which we find in human societies. The underlying model of cooperation has been described in [15, 19]. Agents are able to recognise so-called *patterns of cooperation* which describe situations where certain types of cooperation are both applicable and suitable. The execution of these patterns is described by cooperation protocols based on speech-acts. Agreements between companies (for example as to the price for delivering an order) are reached by negotiation.

If we try to describe the HC setting in terms of task decomposition, it implements a decentralised model of task decomposition: the transportation companies negotiate on how they might decompose the transportation orders. In a certain sense, task allocation is decentralised, too, since the companies generate proposals how to allocate the orders among their group.

Enhanced Cooperation (EC) In the *enhanced cooperation* setting, the relationship among the companies is the same as in HC in a sense that they can exchange orders and information about free capacities. What is new is that it extends the vertical cooperation. Trucks are assigned the ability to reflect on their plans. If a truck agent realises that it has a poor plan, it can cancel the contract made with its company for a specific order, i.e. it may give back its part of that order to its company. In this case, the order is offered again to the agent society. This procedure makes sure that the order will be executed in any case: if the society does not find a better way to process the order, the order automatically falls back to the truck (here, of course, infinite loops and cycles must be avoided by checking appropriate conditions). Thus, there is some risk in following the EC strategy: if no other company can do the task better, the truck has gained nothing, but only lost time.

EC can be regarded as a decentralisation of the hierarchical task allocation implemented by the VC setting. According to EC, the trucks do not always and unconditionally have to accept a task allocation proposed by their company, but can undo this decision in certain cases.

3.3 Task Decomposition and Task Allocation in MARS

The development of multi-agent systems like MARS is motivated by the goal of providing a special purpose system that is able to accomplish a certain set of tasks. In the transportation domain these tasks are the transportation orders given by the users. An important question for the modelling of this domain is how these orders are allocated to the local resources of the companies, namely to their trucks.

In section 2.1, it was already mentioned that the complexity of an order may exceed the capacity of a single shipping company. Therefore, the handling of an order consists of two phases: the *task decomposition phase* and the *task allocation phase*. During the former phase, a decomposition of a task (or an order) into a set of subtasks is computed recursively, until every subtask is small enough to be directly given to some truck. In the latter phase, particular agents have to be determined who will commit themselves to accomplishing the task. In general, each of these phases can be implemented in either a centralised or a decentralised manner, yielding four possible methods for task handling.

Within our current implementation of the scenario at least three of these possible approaches may be found on different levels of the task handling process: the *centralised task decomposition model* (contract-net model), the *decentralised task decomposition model*, and the *completely decentralised model*. The first one, which is characterised by centralised decomposition and centralised allocation, is mainly used between a shipping company and its truck. It is implemented according to a contract net protocol (cf. [8]) as described in chapter 3.1.

In general, the division of a task into a set of subtasks will be done by some heuristic that is available to the decomposition process. But as we are going to deal with open systems in the sense of [13], the heuristics *cannot* take care of the situation of all the agents that are currently part of the system. Moreover, in [16] we presented an example that the decomposition using a heuristic that is not adequate at the moment may fail to compute a solution to a decomposition problem, although an appropriate decomposition is obvious from a more global point of view. To overcome this, we proposed a decentralisation of the task decomposition process: Instead of offering subtasks to the trucks, the whole transportation order is offered to them. Their bids now consist of two parts: firstly, a part of the order they are able to accomplish, and secondly, an estimation of the cost associated with this partial order. The shipping company collects the proposals given by the trucks and uses this information to allocate the subtasks.

In the completely decentralised model, task decomposition as well as task allocation phase are implemented as decentralised processes. It is used e.g. to describe cooperation between autonomous shipping companies, e.g. for dealing with a transportation order that exceeds the capacities of a single company. A company announces its interest in such an order to other companies and specifies a possible subtask of the order she would like to accomplish. The other companies may respond in the same manner, i.e. by announcing their interest in another possible subtask of this order, or they may respond by proposing a modification of the actual state of task decomposition and task allocation. The central technique for achieving a task decomposition and a task allocation that is commonly accepted is the negotiation of the different proposals among the companies that are involved in this process. This process was described in more detail in [16].

In the following section, we present an example and discuss how these different methods can influence the final solution.

4 Cooperation Settings: an Example

In this section, by means of the example shown in figure 1, we explain how vertical (VC), horizontal (HC), and enhanced (EC) cooperation work in practice: we consider two shipping companies, S_1 and S_2. S_1 resides in city B and S_2 in city D. Each shipping company controls a set of two trucks each of which has a capacity of 40 units. The map contains 6 cities, named A, B, C, D, E, and F. The distances between the cities are given by the table shown in figure 1. Let us assume that the system receives a set of orders in the following sequence:

O_1 : 50 units from city A to city E; offered to S_1
O_2 : 10 units from city D to city E; offered to S_1
O_3 : 20 units from city D to city F; offered to S_2
O_4 : 20 units from city E to city C; offered to S_1
O_5 : 20 units from city E to city C; offered to S_1

Example 1: Vertical Cooperation: In this first example, the shipping companies try to solve the problems on their own. The solution in this example is produced using purely vertical cooperation (VC) between the companies and their trucks. In doing so, truck $T_1^{S_1}$ of shipping company S_1 starts from city B to city A to collect 40 units of order O_1, it then goes down to city E to drop 40 units of order O_1. It collects the orders O_4 and O_5 to bring them to city C. Another truck $T_2^{S_1}$

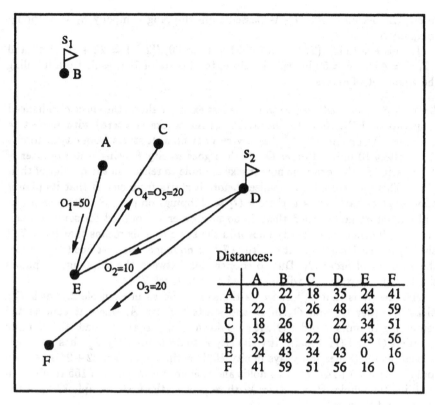

Distances:

	A	B	C	D	E	F
A	0	22	18	35	24	41
B	22	0	26	48	43	59
C	18	26	0	22	34	51
D	35	48	22	0	43	56
E	24	43	34	43	0	16
F	41	59	51	56	16	0

Fig. 1. Example Scenario.

starts from city B to city A to collect the 10 units which were left by truck $T_1^{S_1}$. It then heads for city D to collect O_2 and goes to city E to drop the 20 units of orders O_1 and order O_2. Truck $T_1^{S_2}$ transports order O_3 from city D to city F.

Let $l(T)$ denote the length of the plan which was executed by truck T. $l(T)$ specifies the costs which were necessary for truck T to fulfil its task. We get $l(T_1^{S_1}) = 22 + 24 + 34 = 80$, $l(T_2^{S_1}) = 22 + 35 + 43 = 100$, and $l(T_1^{S_2}) = 56$. Therefore, total costs of 236 were necessary to fulfil the whole set of orders. It it easy to see that, from a global point of view, this solution is not very good. But in fact, it is an optimal solution from the local point of view of each shipping company.

Example 2: Horizontal Cooperation In this example everything remains the same as in example 1 except that the shipping companies do cooperate by offering orders to each other (see section 3.2). In our example, an offer made by another company is accepted if it is better than the offers made by the own trucks. If the shipping companies do behave like this, the solution for truck $T_1^{S_1}$ is the same as in example 1. Also, truck $T_2^{S_1}$ still starts from city B to city A to collect the 10 units of order O_1 which were left by $T_1^{S_1}$. This time, however, it heads directly to city E, because order O_2 is passed from shipping company S_1 to shipping company S_2.

In this example, $l(T_1^{S_1}) = 22 + 24 + 34 = 80$, $l(T_2^{S_1}) = 22 + 24 = 46$ and $l(T_1^{S_2}) = 43 + 16 = 59$ holds. Therefore, total costs of 185 result from fulfilling the whole set of orders.

Example 3: Enhanced Cooperation The last example shows the effect of enhanced cooperation (EC). Here, the behaviour of the trucks is altered with respect to example 2. In example 2, $T_2^{S_1}$ has a very poor plan: It starts from city B to city A to collect 10 units of order O_1. It then goes to city E using just a quarter of its capacity. In this example now trucks are able to reflect on the quality of their plans. This means that $T_2^{S_1}$ realises before it starts from city B that its plan is poor. It gives back its part of order O_1 to shipping company S_1. Because trucks will only start going when there is no new order announced for some amount of time, all orders are already known in the system. This means that truck $T_1^{S_2}$ already has its local plan. At this time $T_2^{S_1}$ is no longer the best one to transport the 10 units of order O_1. Due to cooperation between the shipping companies the 10 units of order O_1 will be passed over to truck $T_1^{S_2}$.

The result is that $T_1^{S_1}$ executes the same plan as in example 2. Truck $T_1^{S_2}$ picks up order O_2 and O_3 in D and starts for city A, where it collects the remaining 10 units of order O_1. It goes down to city E and unloads the 20 units of orders O_1 and O_2. Finally, it visits city F to drop order O_3. $T_2^{S_1}$ has no longer a local plan and therefore stays in city B. This time, $l(T_1^{S_1}) = 22 + 24 + 34 = 80$ and $l(T_1^{S_2}) = 35 + 24 + 16 = 75$ holds and therefore total costs of 155 result from fulfilling the whole set of orders. With respect to the first example 34 % of the costs could be saved.

The above examples can be demonstrated by our implementation of a multi-agent system for the transportation domain. The question is whether these nice problem-solving strategies will work in practice, too. Section 5 describes a series of experiments, where examples of 50 and 400 orders are simulated and evaluated.

5 Experimental Results

In this section, we describe a number of experiments we have run with the MARS system. As mentioned before, the main question is how different kinds of cooperation bring about different solutions to the task decomposition problem, and thus, lead to different behaviour of the system as a whole. In the previous section, three interesting cooperation settings have been discussed, namely VC, HC, and EC. We have seen how these settings correspond to the models of task decomposition and allocation presented in section 3.3. In the following, we will evaluate the different settings by means of a series of experiments.

5.1 Description of the Experiment

In our experiment, three transportation companies with their trucks had to carry out a number of orders. The following parameters could be varied:

- The number of trucks per company varied from 1 to 20.
- The number of orders varied: we tested the system with order loads of 50 and of 400 orders. Moreover, in order to ensure general validity of the results, each experiment was repeated 10 times with randomly generated loads.
- Each experiment was carried out for the VC, the HC, and the EC setting.

The experiments were run on a network of SUN SPARC stations. For the biggest experiment, the agents were run on eight SUN workstations in parallel. What was measured in the experiments were the costs caused by the trucks for carrying out the orders, as well as the average percentage of capacity load of the trucks. The costs were computed by a simple cost model: the costs caused by a truck are proportional to the distance covered while carrying out the orders.

5.2 Results

Figure 2 shows the result of *experiment 1* (which actually has been a series of experiments), which was run with an order load of 50 orders. Figure (2.1) displays the cost caused for solving the scheduling problem. Figure (2.2) shows the average capacity load of the trucks. The x-axis denotes the number of trucks per company. The y-axis is labelled with the cost and the load percentage, respectively. The curves show the behaviour of the system for the three cooperation

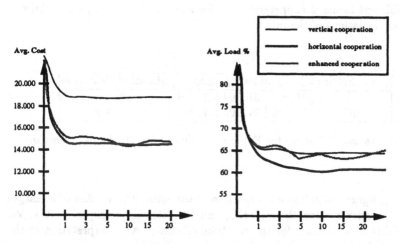

Fig. 2. Experiment 1

settings defined in section 4.

Experiment 2 is a reiteration of the former experiments with a bigger order load. This time, the society of transportation companies had to deliver a set of 400 orders. Figure 3 reveals the results for this scenario. Again, in figure (3.1), the costs are displayed, whereas in figure (3.2), the average capacity load of the trucks is shown.

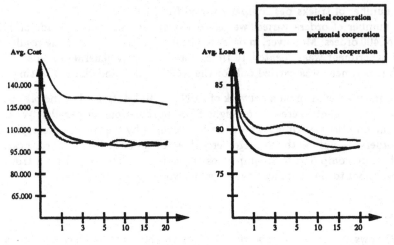

Fig. 3. Experiment 2

In figure 4, some statistical indices characterising the solutions found by VC and HC for experiment one with a number of five trucks per company are shown. Note, that the standard deviation can be regarded as a measure for the stability of the different forms of cooperation, i.e. for how much the computed solutions depend on variations of the input.

Coop. Setting	$Cost_{avg}$	$Cost_{max}$	$Cost_{min}$	Standard Deviation
VC	16,859	31,604	7,706	5677.6
HC	13,365	21,200	7,210	3560.0

Fig. 4. Statistical Indices for Experiment One, Five Trucks per Company

Finally, figure 5 displays a comparison of the estimated number of messages sent by the agents using the VC, HC, and EC settings for experiment 1. We will use this as a measure for the run-time efficiency. An interpretation of the experimental data is provided in the following subsection.

5.3 Discussion

Let us now have a closer look at the basic results of the experiments described in the previous section.

Quality of the Solutions: If we compare the costs in figures (2.1) and (3.1), the most obvious result is that in both experiments, the average cost can be reduced

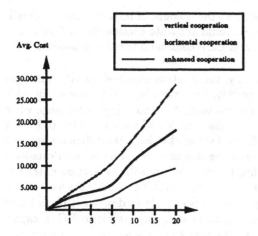

Fig. 5. Message Statistics for Experiment 1

considerably by introducing horizontal cooperation. Moreover, as problems get more complex, using horizontal cooperation pays off more and more: if we look at the case of 20 trucks, using the HC setting in the experiment 1 reduces the cost by about 21% compared to the VC solution, whereas in the bigger experiment 2, cost are reduced by 28%. On the other hand, the EC strategy does not seem to yield considerably better results than HC. This is discussed below.

As regards the average capacity load of the trucks, there are three remarkable points: Firstly, the VC strategy yields a slightly better load percentage than the HC strategy. Secondly, with the number of trucks increasing, the load percentage tends to decrease. Thirdly, the EC strategy yields the best capacity loads, on an average. What seems to be a bit confusing at a first glance is that the poor VC strategy results in better capacity loads than the sophisticated HC. Intuitively, we could think that lower costs and a better utilisation of capacity go hand in hand. In fact, the relationship is not as straightforward as it seems to be. A truck can go long ways - as long as it is fully loaded, the capacity load will be ok. Therefore, a single truck can reach a high capacity load by combining orders in a clever manner - this, however, does not automatically imply that the plan causes little costs. This is the main reason why VC results in a good capacity usage. On the other hand, EC combines the low cost of HC with the good capacity load of VC.

Runtime Efficiency: Run-time efficiency was measured by the number of messages which were sent by agents to other agents. Here, the results clearly confirmed our assumptions: a higher communication overhead is the price to pay for the better solutions obtained by using the EC and HC settings. However, this overhead drastically depends both on the internal decision criteria of the companies and on the negotiation protocols used. In the example, each company offers each order to any acquainted other company. By using more intelligent (heuristic) criteria for partner selection, the amount of messages can be drastically reduced. In our experiments, we could not state considerable differences in run-time between VC and HC, whereas EC is considerably more time-consuming.

Stability and Convergence: Obviously, the solution obtained by HC converges against a stable local state very quickly. For example, in experiment one, this stable state is reached with only five trucks, and the cost does not change when more trucks are used. Moreover, the behaviour of HC systems seems to be less sensitive to changing input data: figure 4 shows that both the difference between minimal and maximal cost values and the standard deviation is much smaller if HC is used than if VC is used. This, however, satisfies the evaluation criteria of graceful degradation and flexibility proposed by [24]. EC is both the least stable and the least predictable strategy, since trucks decide to drop their plans based solely on local quality criteria. EC produces cost comparable to HC, a capacity usage comparable to VC, but a very high computation and communication overhead. Thus, the experience is confirmed that modelling in the small requires one to be very careful, since locally reasonable decisions lead to factually no global improvement, or even to a decrease of the system behaviour, unless the design and the parameter setting are done in a very cautious way. Finding better decision criteria for EC is a subject of our future work.

Optimal Solution Up to now, we have not said anything about optimal solutions to the scheduling problems solved by our system in the two experiments. Of course, it would be fine to know about the real optima. However, since the orders arrive asynchronously and are scheduled dynamically, and since message deliverance times have to be taken into consideration, such a reference solution cannot be obtained by using static OR methods like Branch and Bound algorithms (see e.g. [5]). Up to now, and as far as we know, no practicable approaches towards solving this problem exist. This corresponds to the fact that evaluation of large distributed systems is a big problem in general.

6 Conclusion

The transportation domain was introduced as a multi agent application, The use of specific DAI techniques for the cooperation between the agents and for task decomposition and allocation were described. By means of a series of experiments, first results as regards the influence of local criteria to the emerging functionality of the system as a whole were reported.

Thus, there are three main contributions of the paper: Firstly, we showed how different models and mechanisms of task decomposition, task allocation, and of cooperation can be used in order to model a real-world application and in order to solve hard real-world problems such as the scheduling problem by applying the paradigm of emergent functionality. Secondly, our empirical results confirm that the multi-agent approach is a *suitable* approach for modelling and solving this kind of problems. Thirdly, our experiments can be useful in a more practical sense: cooperation among transportation companies appears to be an important subject when it comes to solve today's world-wide traffic problems. The reduction of costs which has been observed in our experiments can turn out to be a strong and convincing argument in favour of this type of cooperation.

References

1. L. Bodin, B. Golden, A. Assad, and M. Ball. Routing and scheduling of vehicles and crews. *Computers and Operations Research*, 10(2):63 – 211, 1983.
2. M. Buchheit, N. Kuhn, J. P. Müller, and M. Pischel. MARS: Modelling a Multiagent Scenario for Shipping Companies. In *Proceedings of the European Simulation Symposium (ESS-92)*, Dresden, 1992. Society for Computer Simulation (SCS).
3. S. Bussmann and H. J. Müller. A Negotiation Framework for Cooperating Agents. In *Proc. of the 2nd Workshop on Cooperating Knowledge Based Systems*. Keele, GB, September 1992.
4. P. Bagchi and B. Nag. Dynamic Vehicle Scheduling: An Expert System Approach. *Journal of Physical Distribution and Logistics Management*, 21(2), 1991.
5. M. Desrochers, J. Desrosiers, and M. Solomon. A new optimisation algorithm for the vehicle routing problem with time windows. *Operations Research*, 40(2), 1992.
6. E. H. Durfee and V. R. Lesser. Negotiating task decomposition and allocation using partial global planning. In *Distributed Artificial Intelligence, Volume II*, pages 229–244, San Mateo, CA, 1989. Morgan Kaufmann Publishers, Inc.
7. W. Domschke. *Logistik (Bd.2): Rundreisen und Touren.* Oldenbourg-Verlag, München - Wien, 3rd edition, 1990.
8. R. Davis and R.G. Smith. Negotiation as a methaphor for distributed problem solving. *Artificail Intelligence*, 20:63 – 109, 1983.
9. K. Fischer and N. Kuhn. A DAI Approach to Modelling the Transportation Domain. Technical Report RR-93-25, DFKI, 1993.
10. K. Fischer, N. Kuhn, and J. P. Müller. Distributed, knowledge-based, reactive scheduling in the transportation domain. In *Proc. of the Tenth IEEE Conference on Artificial Intelligence and Applications*, San Antonio, Texas, March 1994.
11. K. Fischer and H. M. Windisch. MAGSY- Ein regelbasiertes Multi-Agentensystem. In H. J. Müller, editor, *KI1/92, Themenheft Verteilte KI*. FBO-Verlag, 1992.
12. M. R. Garey and D. S. Johnson. *Computers and Intractability - A Guide to the Theory of NP-Completeness.* W.H. Freeman, 1979.
13. C. Hewitt. Open information systems semantics for distributed artificial intelligence. *Artificial Intelligence*, 47:79–106, 1991.
14. N. R. Jennings. *Joint Intentions as a Model of Multi-Agent Cooperation.* PhD thesis, Queen Mary and Westfield College, London, August 1992.
15. N. Kuhn, H. J. Müller, and J. P. Müller. Simulating cooperative transportation companies. In *Proceedings of the European Simulation Multiconference (ESM-93)*, Lyon, France, June 1993. Society for Computer Simulation.
16. N. Kuhn, H. J. Müller, and J. P. Müller. Task decomposition in dynamic agent societies. In *Proceedings of the International Symposium on Autonomous Decentralised Systems (ISADS-93)*, Tokyo, Japan, 1993. IEEE Computer Society Press.
17. S. E. Lander and V. R. Lesser. Understanding the role of negotiation in distributed search among heterogeneous agents. In *Proc. of the 12th International Workshop on*

Distributed Artificial Intelligence, pages 249–262, Hidden Valley, Pennsylvania, May 1993.

18. H. Müller-Merbach. *Operations Research*. Verlag Franz Vahlen, München, 3rd edition, 1973.

19. J. P. Müller and M. Pischel. The Agent Architecture INTERRAP: Concept and Application. Technical Report RR-93-26, German Artificial Intelligence Research Center (DFKI), Saarbrücken, June 1993.

20. J. P. Müller and M. Pischel. Integrating agent interaction into a planner-reactor architecture. In *Proc. of the 13th International Workshop on Distributed Artificial Intelligence*, Seattle, WA, USA, July 1994.

21. J. P. Müller and M. Pischel. Modelling interacting agents in dynamic environments. In *Proc. of the European Conference on Artificial Intelligence (ECAI94)*, pages 709–713. John Wiley and Sons, August 1994.

22. H. Van Dyke Parunak. Industrial applications of multi-agent systems. In *Proc. of INFAUTOM-93*, Toulouse, February 1993. Association Colloque SUP'AERO.

23. R. Rittmann. Die Macht der Trucks. *Bild der Wissenschaft*, 9:112–114, 1991.

24. L. Steels. Cooperation between distributed agents through self-organisation. In Y. Demazeau and J.-P. Müller, editors, *Decentralised A.I.*, pages 175–196. North-Holland, 1990.

25. K. P. Sycara. Multiagent compromise via negotiation. In L. Gasser and M. N. Huhns, editors, *Distributed Artificial Intelligence, Volume II*, pages 119–137. Morgan Kaufmann, San Mateo, California, 1989.

26. H.-J. Zimmermann. *Methoden und Modelle des Operations Research*. Vieweg-Verlag, Braunschweig - Wiesbaden, 2nd edition, 1992.

27. G. Zlotkin and J. S. Rosenschein. Negotiation and task sharing among autonomous agents in cooperative domains. In *Proc. of the Eleventh IJCAI*, pages 912–917. Detroit, Michigan, August 1989.

A Framework for the Interleaving of Execution and Planning for Dynamic Tasks by Multiple Agents

Eithan Ephrati[1] and Jeffrey S. Rosenschein[2]

[1] Computer Science Department
University of Pittsburgh
Pittsburgh, PA
tantush@cs.pitt.edu
[2] Institute of Computer Science
The Hebrew University
Jerusalem, Israel
jeff@cs.huji.ac.il

Abstract. The subject of multi-agent planning has been of continuing concern in Distributed Artificial Intelligence (DAI). In this paper, we suggest an approach to the interleaving of execution and planning for dynamic tasks by groups of multiple agents. Agents are dynamically assigned individual tasks that together achieve some dynamically changing global goal. Each agent solves (constructs the plan for) its individual task, then the local plans are merged to determine the next activity step of the entire group in its attempt to accomplish the global goal. Individual tasks may be changed during execution (due to changes in the global goal).

The suggested approach reduces overall planning time and derives a plan that approximates the optimal global plan that would have been derived by a central planner.

1 Introduction

The subject of multi-agent planning has been of continuing concern in Distributed Artificial Intelligence (DAI). Some of the earliest work in the field, such as Smith's Contract Net [15] and Corkill's work on distributing the NOAH planner [2], directly addressed the question of how to get groups of agents to carry out coordinated, coherent activity. These questions have occupied researchers both in the Cooperative Problem Solving subarea of DAI (researchers concerned with centrally designed groups of agents with a global goal), as well as those working on Multi-Agent Systems (collections of self-interested agents with no global goal). More recent research on the issue of multi-agent planning includes that of Martial [19, 18], Katz and Rosenschein [10], Durfee [3, 5, 4], and Lansky [12].

The term multi-agent planning has been used to describe both "planning *for* multiple agents" (where the planning process itself may be centralised), and "planning *by* multiple agents" (where the planning process is distributed among the agents). In this paper, we are concerned with the latter paradigm:

we investigate how a group of agents can cooperatively construct a dynamically changing plan which, as a group, they also carry out.

We present a multi-agent planning procedure to achieve a dynamic global goal; that is, the global goal itself may change over time. The procedure relies on the dynamic distribution of individual planning tasks. Agents solve their local tasks, which are then merged into a global plan. By making use of the computational power of multiple agents working in parallel, the process is able to reduce the total elapsed time for planning and is better able to deal with changes in objective as compared to a central planner.

2 The Scenario

Our scenario involves a group $A = \{a_1, \ldots, a_m\}$ of m agents. These agents are to achieve at any given time (t) a global goal (G^t, which may change over time). Through the distribution of individual planning tasks ($\{g_1^t, \ldots, g_n^t\}$) each agent determines p_i^t, the plan (expressed as a set of constraints) that accomplishes its present task(s). Then, based on the individual plans, the agents communicate as they construct the next multiagent step towards the achievement of the global goal. During this interleaved process of planning and execution, some agents may be assigned different tasks. For the time being, we assume that the agents are benevolent and cooperative; however, see Section 5.

2.1 Assumptions and Definitions

- The *global goal* at time t, G^t, is a set of predicates, possibly including uninstantiated variables. G^t is divided [3] into n tasks $[g_1^t, g_2^t, \ldots, g_n^t]$. Each task is a set of instantiated (grounded) predicates such that their union satisfies G^t ($\cup_i g_i^t \models G^t$). (Tasks are assigned to individual agents only for the purpose of the planning process; the actual execution is divided differently.)
- Any plan that achieves g_i^t is denoted by p_i. Each such sub-plan is a sequence of operators $\langle op_1^t, \ldots, op_k^t \rangle$ that, invoked in the current configuration of the world S^t, establishes g_i^t.
- Each agent has a *cost function* ($C \colon OP \times A \Rightarrow \mathbb{R}$) over the domain's operators [8]. The cost of a_j's plan $c_j(p_j)$ is defined to be $\sum_{k=1}^m c_j(op_k)$.

3 An Example

As an example, consider a simple scenario in the slotted blocks world as described in Figure 1.[4] There are two agents (a_1, a_2) and 3 blocks (a,b,c) with a length

---

[3] Division of tasks is done by a coordinator or a "supervisor" who is in charge of the group of agents.

[4] While the blocks world is inappropriate for studying many real-world issues in robotics, it remains suitable for the study of abstract goal interactions.

of 2 feet each. The ground surface is divided into 30 contiguous marked regions (coordinates) of 1 foot each, aligned in a row.

The world may be described by the following relations: **Clear**(b)—there is no object on b; **On**$(b, x, V/H)$—b is located on block x (or at location x) either vertically (V) or horizontally (H); **At**(x, loc)—the left edge of object x (agent or block) is at loc. The functions $r(b)$ and $l(b)$ return the grid region of b's left edge, and the length of b, respectively. From now on we will use only the first letter of a predicate to denote it.

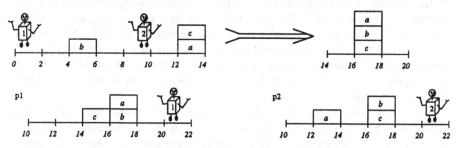

Fig. 1. A Simple Sticky Blocks World Example

The initial state is:
$\{A(a_1, 0), A(a_2, 9), A(a, 12), O(a, 12, H), A(c, 12), O(c, a, H), A(b, 4), O(b, 4, H)\}$.

The available operators (described in a STRIPS-like fashion) are:[5]

Take$_i(b, x, y)$— take b from x to y (x and y denote either a region or another block).

    [cost: $|loc(x) - loc(y)| \times l(b)$,
    prec: $O(b, x, z), C(b), C(y), A(a_i, x), A(b, x)$
    del: $C(y), A(a_i, x), A(b, x)$,
    add: $O(b, y, z), A(a_i, y), A(b, y)]$

Move$_i(a_i, x, y)$—Go from x to y.

    [cost: $|x - y|$,
    prec: $A(a_i, x)$,
    del: $A(a_i, x)$,
    add: $A(a_i, y)]$

The agent's initial tasks are (respectively):
    $g_1^0 = \{A(a, 16), A(b, 16), O(a, b, H)\}$ and
    $g_2^0 = \{A(b, 16), A(c, 16), O(b, c, H)\}$.

The combination of these two tasks will eventually establish the global goal $G^0 = \{A(a, 16), A(b, 16), O(a, b, H)\}$.[6] The lower part of Figure 1 describes the emergent state of each task if invoked alone.

---

[5] We assume in this specific example that all agents have identical capabilities. The operators are indexed with respect to the agent that performs them.

[6] This is reminiscent of Sussman's Anomaly in the single-agent planning scenario— where the plan to achieve one subgoal obstructs the plan that achieves the other [16].

After 3 execution time-steps the task g_1^0 is change to be $\{A(a, 16), O(a, c, H)\}$ ($= g_1^3$) and g_2^0 is cancelled.

Given these (changing) tasks, our agents are to go through a planning process that will result in satisfying the final goal.

4   General Overview of the Main Process

In this section, we consider the primary phase of our technique, namely merging the sub-plans that achieve the given task so as to determine the next optimal move of the entire group. A sub-plan is constructed by an agent with only a local view of the overall problem; therefore, conflicts may exist among agents' sub-plans, and redundant actions may also have been generated. Given the set of sub-plans, we are looking for a method to merge them in an optimal (and inexpensive) way.

In this section, we describe a cost-driven merging process that results in a coherent global plan (of which the first set of simultaneous operators is most relevant), given the sub-plans. We assume that at each time step t each agent, i, has been assigned (only for the purposes of the *planning process*) one task. Given that instance, the agent will derive p_i^t, the sub-plan that achieves it. Note that once i has been assigned g_i^t at any given t, the plan it derives to accomplish it stays valid (for the use of the algorithm) as long as g_i^t remains the same. That is, for any time $t + k$ such that $g_i^{t+k} = g_i^t$, it holds that $p_i^{t+k} = p_i^t$. Therefore, replanning can be modularised among agents; one agent may have to replan, but the others can remain with their previous plans.

The process is iterative. The underlying idea is the *dynamic generation of alternative execution steps* that identifies the optimal global plan and thus the next optimal global execution step. At each step, all agents state additional information about the sub-plan of which they are in charge. The next optimal step is then determined and the current configuration of the world, S^t, is changed to be S^{t+1}. The process continues until all tasks have been accomplished (the global goal as of that specific time has been achieved). Plans are represented as sets of absolutely necessary sets of constraints that enable them.

Note that when we refer to an individual agent's sub-plan, we are really only concerned with that initial part of the sub-plan that is relevant to the merging process at any specific time step t (which, in turn, might depend on some "look-ahead factor," how many steps of the plan are computed prior to actual execution of a plan step). The way each individual determines this initial part of his sub-plan (i.e., what to do next) under time constraints is beyond the scope of this paper.

Essentially, the search method employs a Hill-Climbing algorithm. The heuristic function (h') that guides the search is dynamically determined by the agents during the process. Actually, h' is the sum of the approximate remaining costs that each agent assigns to his own sub-plan. Note that this heuristic function is powerful enough to avoid the classic problems of Hill Climbing (foothills, plateaus, and ridges) since at each time step it assures some progress in the right direction.

In general h' is an underestimate, since plans will tend to interfere with one another, but due to overlapping constraints ("favor relations" [19], or "positive" interactions) it might sometimes be an overestimate.

4.1 More Definitions

Having presented the broad outlines of the merging process, we now introduce some additional definitions.

- $\hat{e}^t_i(g^t_i)$ is the set of absolutely necessary general constraints needed for any plan to achieve the grounded instance of the subgoal g^t_i, starting with some initial state S^t. In accordance with the partial order over these constraints,[7] we divide $\hat{e}^t_i(g^t_i)$ into subsets of constraints. Each such subset within $\hat{e}^t_i(g^t_i)$ comprises all the constraints that can be satisfied within j (optimal) steps, and are necessary at some subsequent step after j. The total number of these subsets is denoted by $\mid \hat{e}^t_i(g^t_i) \mid$.

- A refinement of $\hat{e}^t_i(g^t_i)$ is either a different temporal order of that set that would still achieve the task, or a further specification (extended partition) of its sub-sets that enable them (the sub-sets). $\mathcal{E}^t_i(g^t_i)$ denotes the set of all the refinements of $\hat{e}^t_i(g^t_i)$. .

We denote the components of $e(g)$ (any member of $\mathcal{E}(g)$) by $\bigcup_j E^j_g$, such that E^j includes all the constraints that can be satisfied within j steps, and are necessary at some step $\geq j$. For any $j \geq \mid e(g) \mid$, we define E^j to be the description of the goal g.

Example: Assume in the slotted blocks world scenario from Section 3, that g^0_1 is determined to be $O(b, a, H), A(b, 13), A(a, 13)$. Then $\hat{e}^0_1(g^0_1)$ for Agent 1 would be: $\hat{e}^0_1 = \{[C(b), C(c)](= \hat{E}^0_1) \cup [A(a_j, r(c)), C(b), C(c)](= \hat{E}^0_1) \cup [C(b), C(a)](= \hat{E}^0_1) \cup [A(a_i, r(a)), C(b), C(a)] \cup [A(a, 13)), O(a, 13, H), C(b), C(a)] \cup [A(a, 13)), O(a, 13, H), C(b), C(a), A(a_i, r(b))] \cup [A(a, 13)), O(a, 13, H), A(b, 13), O(b, a, H)]\}$ (inducing the plan $\langle M(r(a_i), r(c)), T(c, 12, 10), M(10, 12), T(a, 12, 13), M(13, 4), T(b, 4, 13))\rangle$.

One refinement of this plan would replace $\hat{E}^0_1$ with $[A(a_j, r(b)), C(b), C(c)] \cup [A(b, 10), C(b), C(c)] \cup [A(a_j, r(c)), C(b), C(c)]$ (inducing the plan: $\langle M(r(a_i), r(b)), T(b, 4, 10), M(r(a_i), r(c)), T(c, 12, 15), M(15, 12), T(a, 12, 13), M(13, 10), T(b, 10, 13))\rangle$.

- $P(E)$ denotes the set of "grounded" plans (what Chapman calls "complete" plans [1]) that is induced by the set of constraints E.

$F^1_{\text{ollow}}(E)$ is defined to be the set of constraints that can be satisfied by invoking at most one operator, given the initial set of constraints E ($F^1_{\text{ollow}}(E) = \{I \mid \exists op \exists P[op(P(E)) \models I]\}$). Similarly $F^2_{\text{ollow}}(E)$ is the set of constraints

---

[7] Constraints are temporally ordered sets of the domain's predicates associated with the appropriate limitations on their codesignation [1].

that can be satisfied by invoking at most two operators *simultaneously* (by two agents) given E, and $F_{ollow}^m(E)$ is the set that can be achieved by m simultaneous actions.

- $p_1 \parallel p_2$ denotes the concatenation of the plans p_1 and p_2. Operators that are invoked simultaneously in a multi-agent plan are grouped together. The order within this group implies the order in which they are completed (i.e., $\langle Op_1, Op_2, \ldots \{Op_j, Op_k\}, \ldots, Op_m \rangle$ denotes the fact that Op_k and Op_j are initiated simultaneously and Op_j will be terminated no later than Op_k).

4.2 The Algorithm

This section describes the algorithm in more detail, along with a running example. At each step of the procedure, agents try to impose more of their private constraints on the group's aggregated set of constraints that has induced the plan which was executed thus far.

The set of constraints that has induced the multi-agent plan that has been executed up to time step t, P^t, is denoted by $\mathcal{A}^t$. $\mathcal{A}^{t+}$ denotes the set of all aggregated sets of constraints at step t.

As an example, consider again the scenario presented in Section 3. To simplify things we will use throughout this example only $\hat{e}_i$ instead of $\mathcal{E}_i$, and in most cases $F_{ollow}^1(\mathcal{A}^{t+})$ instead of $F_{ollow}^m(\mathcal{A}^{t+})$.

The exact procedure is defined as follows:

1. At step 0 each agent i finds $\mathcal{E}_i^0(g_i)$—the set of all alternative absolutely necessary temporally ordered sets of constraints that achieve the task g_i^0, starting from the initial state S^0, and its refinements. Actually, the agent has to determine only the the first l subsets of this set depending on the look-ahead factor.

   The set of alternatives is initialised to be the empty set $(\mathcal{A}^0 = \emptyset)$.

   In our example, we have: $\hat{e}_1^0(g_1) = \{[C(b), C(c)] \cup [A(a_i, r(b)), C(b), C(c)] \cup [C(b), C(c), A(b, 16)] \cup [A(a_i, r(c)), C(b), C(c), A(b, 16)] \cup [C(a), C(b), A(b, 16)] \cup [A(a_i, r(a), C(a), C(b), A(b, 16)] \cup [O(a, b, H), A(a, 16), A(b, 16)]\}$

   (this ordered set induces the plan $\langle M(0, 4), T(b, 4, 16), M(16, 12), T(c, 12, x_c = 14 \mid 12), M(x_c, 12), T(a, 12, 16)\rangle$)

   and $\hat{e}_2^0(g_2) = \{[C(c), C(b)] \cup [A(a_j, r(c), C(c), C(b)] \cup [C(b), A(c, 16), C(c)] \cup [A(a_j, r(b), C(b), A(c, 16), C(c)] \cup [A(b, 16), A(c, 16), O(b, c, H)]\}$

   (inducing the plan $\langle M(9, 12), T(c, 12, 16), M(16, 4), T(b, 4, 16)\rangle$). Note, that $\mathcal{E}_i$ has in general, even in this example, more elements. For example, $\mathcal{E}_1$ would also include a different order, in which block c is removed from the top of block a before block b is placed at region 16, and $\mathcal{E}_2$ would include the refinement of e_2 in which, prior to removing block c, block b is moved to region 14 (inducing a_2's optimal plan: $\langle M(9, 4), T(b, 4, 14), M(14, 12), T(c, 12, 16), M(16, 14), T(c, 14, 16)\rangle$)

2. At step t each agent declares $E_i^{A_j^t} \subseteq E_i^{l+1}$ only if for all $k \leq l$ E_i^t is already modelled by P^t (i.e., all its predecessors in the private sub-plan were already declared and merged into the global plan), and the declaration is "feasible," i.e., it can be reached by invoking at most n operators simultaneously on that set: $(\bigcup_{n=1}^{l} E_i^l \subseteq A_j^t) \wedge (E_i^{A_j^{1+}} \subseteq F_{\text{ollow}}^m(A_j^t))$.

   i can try to contribute elements of his "next" private subset of constraints to the global set only if they are still relevant and his previous constraints were accepted by the group.

   At the first step, there is only one candidate set for extension — the empty set. Each agent i may declare $E_{g_i}^1$. In our example, this will be $E_{g_1}^1 = E_{g_2}^1 = [C(c), C(b)]$.

   At the second step, $A^1 = [C(c), C(b)]$. a_1 declares $E_1^{A_1^{1+}}$, which in this example is equal to $E_{g_1}^2 = [A(a_i, r(b)), C(b), C(c)]$. Similarly, a_2 declares $E_{g_2}^2 = [A(a_j, r(c)), C(c), C(b)]$. (Both are in $F_{\text{ollow}}^1(A^1)$, which contains only one subset.)

   At the third step $A^2 = [A(a_i, r(b)), A(a_j, r(c)), C(c), C(b)]$, a_1 declares $[A(b, 16), C(b), C(c)]$ and a_2 declares $[A(c, 16), C(b), C(c)]$.

   At the fourth step e_1^0 is no longer valid, instead a_1 declares the set $E_1^{A^3} = [A(a_i, r(b)), C(b), A(a_j, r(a)), C(a)]$ (which is in $F_{\text{ollow}}^2(A^3)$) while a_2 has no task at all. At the fifth step a_1 declares $[O(a, c, H), A(a, 16)]$, which satisfies the final goal.

3. At this step, $\{Ex_r(A_j^t)\}$, all the maximal consistent extensions of A_j^t with elements of $\bigcup_i E_i^{A_j^{1+}}$ are generated (where each $Ex(A_j^t)$ is defined as the following fixed point: $\{I \mid (I \in \bigcup_i E_i^{A_j^{1+}}) \wedge (I \cup Ex(A_j^t) \not\models \text{False})\}$).

   At the first step, the aggregated set of constraints is $[C(c), C(b)]$.

   At the second step, both declarations may coexist consistently, and there is therefore only one successor to the previous set of constraints: $A^2 = A^1 \cup [A(a_i, r(b), A(a_j, r(c), C(c), C(b)]$.

   At the third step, A^2 has 4 possible extensions: $Ex_1 = [A(c, 16), C(b), C(c)]$, $Ex_2 = [A(b, 16), C(b), C(c)]$, $Ex_3 = [A(b, 16), C(b), A(c, 16)]$ (where b is located on a), and $Ex_4 = [A(b, 16), C(b), A(c, 16)]$ (where b is located on a). Thus, A^3 has four different members (the union of A^2 with each of the extensions).

   At the fourth step, A^3 has one extension: $[A(a_i, r(b)), C(b), A(a_j, r(a)), C(a)]$ which includes both agents' declarations.

   At the fifth step, the sole extension of A^4 is the description of both tasks at time 4.

4. All extensions are evaluated so as to choose the next step of the multi-agent plan (i.e., find the h value of the Hill-Climbing search). Each agent declares, h_i, the estimate it associates with each newly-formed set of aggregated constraints (the cost of completing "his private" task given that set). The h value is then taken to be the sum of these estimates ($h(Ex_r(A^{t+})) = \sum_i h_i'(Ex_r(A^{t+}))$).

There are two possible policies to determine the h value by each agent. According to the first (the "rigid policy"), each agent would stick to the initial plan that he generated. The h'_i would then be equal to the cost of completing the plan, given the furthest satisfied point of his set of constraints according to that initial plan. A more accurate but also more expensive way (the "interactive policy"), would have the agent reconstruct the plan, given the constraints that have been achieved so far. Note that such a re-planning process will not be too expensive, since from a certain point the new private plan would become identical to the initial private plan. Also note that a more accurate estimate would also take into account the actual cost of achieving each alternative extension.

A^1 is satisfied by the initial state, therefore, $h(A^1)$ is equal to its h value (that is, the sum of the individual estimate costs, which is 89). This value changes at the second step to be 82 since both agents carry out the first step of their initial plans.

At the third step, the four possible extensions score respectively the following h values according to the "rigid policy": $60, 74, 40, 82$. According to the "interactive policy", the scores are $42, 66, 12, 36$. As an example consider $Ex_1(A_1)$: this set fully satisfies $E_1^1 \cup E_1^2 \cup E_1^3$. Therefore, according to both policies, a_1 gives it the heuristic value of 18. But according to a_2's evaluation, only $E_2^1 \cup E_2^2$ are satisfied. Therefore, a_2 declares the estimated cost of achieving his final subgoal, given only this subset of his full set of constraints. According to his original plan, this cost is 42; actually re-planning to achieve its final goal, given A_1^4, would yield the plan
$\langle T(c, 12, 14), M(14, 16), T(b, 16, 18), M(18, 14), T(c, 14, 16), M(16, 18), T(b, 18, 16)\rangle$ which actually costs only 24. Therefore, the third extension is chosen.

At the next two steps only one extension is considered.

5. Each extension induces a sequence of operators that achieves it $(P(Ex_r(A_j^{t+})))$. At this stage, based on the best alternative extension, $Ex_r^*(A_j^{t+})$, that was chosen in the previous step, the agents execute additions to the ongoing executed plan. The generation of these sequences is fully described in Section 4.3.

The corresponding multi-agent plan that has been generated so far is concatenated with the resulting segments of the plan extension:
$P(A_j^t) \parallel \{P(Ex_r^*(A_j^{t+}))\}$. Similarly, the aggregated set A_j^t is replaced by its union with its extensions: $A^t + 1 = (A^t \cup \{A_j^{t+} \cup Ex_r(A_j^{t+})\}$.

At the first step the (sole) extension is fully satisfied by the initial state; therefore, $g(A^1) = 0$. At the second step, $Ex(A^1)$ can be achieved by $\langle\{M_1(0, 4), M_2(9, 12)\}\rangle$. The four possible extensions that are generated in the third step can be achieved (respectively) by the following plans $\langle T_1(b, 4, 16)\rangle$, $\langle T_2(c, 12, 16)\rangle$, $\langle\{T_1(b, 4, 16), T_2(c, 12, 16)\}\rangle$; and $\langle\{T_2(c, 12, 16), T_1(b, 4, 16)\}\rangle$.
At the fourth step the extension is achieved by $\langle M_1(16, 12), T_2(b, 16, 18)\rangle$. The final extension is established by $\langle T_1(a, 12, 16)\rangle$.

6. The process ends when no further tasks are given.
   In the example we stop the search when the goal is achieved—at the fifth step.

4.3 Construction of the Multi-Agent Plan

The multi-agent plan is constructed and executed throughout the process. The construction is made at Step 5 of the algorithm. At this step, all the optimal sequences of moves are determined. The construction of the new segments of plans is determined by the cost that agents assign to each of the required actions; each agent bids for each action that $Ex_r^*(A_j^{t+})$ implies. The bid is based on the sequence of i's actions in $P(A_j^t)$. Thus, the minimal cost sequence, $P(Ex_r^*(A_j^{t+}))$, is determined.

An important aspect of the process is that each extension of the set of constraints belongs to the F_{ollow}^m of the already achieved set. Therefore, it is straightforward to detect actions that can be taken in parallel. Thus the plan that is constructed is not just cost efficient, but also time efficient. In this framework, we give primary importance to the cost of the resulting global plan, but if the time of execution is of equal importance, it is possible to maximise the utility of the constructed plan (instead of just minimising its cost) where the utility is a function of both execution time and cost.

There is an important tradeoff to be made here in the algorithm. Since the agents choose the least expensive sequence of additional steps, considering them only in isolation, and make the best current decision, the resulting global plan cannot be guaranteed optimal. The effect of this drawback might be reduced by increasing the time spent on planning at the expense of delaying execution (this is, in fact, the "look-ahead factor" mentioned above). Having the agents plan larger segments of the plan before actually executing it will improve the eventual result, but on the other hand will complicate the search, and delay execution (which may be undesirable).

Note that there is another tradeoff to be made here. We let each agent declare $F_{ollow}^m(A^t)$ to ensure optimality of parallelism in the resulting global plan. However, we can relax this demand, so as to have agents relate just to $F_{ollow}^1(A^t)$ and establish only partial parallelism.

In the example all the suggested methods would yield the same final plan. At the first step the (sole) extension is fully satisfied by the initial state. Therefore, no plan is constructed.

At the second step $Ex(A^1) = [A(a_i, r(b)), A(a_j, r(c)), C(c), C(b)]$. These constraints can be achieved by $M_i(r(a_i), r(b))$ and $M_j(r(a_j), r(c))$. The bids that a_1 and a_2 give to these actions are respectively $[4, 12]$ and $[6, 3]$. Therefore, a_1 is "assigned" to block b and a_2 is assigned to block c. The constructed plan is $\langle\{M_1(0, 4), M_2(9, 12)\}\rangle$.

Based on the agents' bids, the chosen extension in the third step can be achieved by: $\langle\{T_1(b, 4, 16), T_2(c, 12, 16)\}\rangle$.

At the fourth step, the extension is achieved by $\langle\{M_i(16, 12), T_j(b, 16, 18)\}\rangle$. Either agent may be assigned each task.

The final extension is established by $\langle T_i(a, 12, 16)\rangle$, and the corresponding final plan that is constructed is: $\langle \{M_1(0, 4), M_2(9, 12)\}, \{T_2(c, 12, 16), T_1(b, 4, 16)\}, \{M_1(16, 12), T_2(b, 16, 18)\}, T_1(a, 12, 16)\rangle$.

Theorem 1. *Let the cost effect of "positive" interactions among members of some set, k, of sub-plans that achieves G be denoted by δ_k^+, and let the cost effect of "negative" interactions among these sub-plans be denoted by δ_k^-. We say that the multi-agent plan that achieves G is δ-optimal, if it diverges from the optimal plan by at most $\max_k |\delta^+ - \delta^-|$.*

Then, at any time step t, employing the merging algorithm, the agent will follow the δ-optimal multi-agent plan that achieves G^t.

Proof. Let P^* be some optimal multiagent plan that achieves G^t within the boundary $|\delta^+ - \delta^-|$, and let $e(G^t)$ denote the set of constraints that induces P^*. Since given $e(G^t)$ the optimal set of corresponding operations is found (Step 5 of the process), it is sufficient to prove that $e(G^t)$ is generated by the process.

The effect of heuristic overestimate (due to positive future interaction between individual plans) and the effect of heuristic underestimate (due to interference between individual plans) balance each other. Therefore, by allowing a deviation of the plan from the optimal one within the boundary of the combined effect, we can consider the heuristic function to be accurate.

The proof is by induction on the subsets of constraints, $e^1(G) \cup e^2(G) \cup \ldots \cup e^n(G)$, that construct $e(G)$ (which in effect equals the number of simultaneous operations that construct P^*).

1. $e^1(G^t)$; $e^1(G^t)$ can be achieved from the initial state by simultaneously invoking $r \le m$ operators. Therefore, each element of $e^1(G^t)$ belongs to the F_{ollow}^m set of the initial state. On the other hand, $e^1(G^t)$ must be constructed out of elements of $e_i^1(g_i) \in \mathcal{E}_i(g_i)$ for some a_i, and each of the corresponding agents, a_i, will be able to declare in the first step all elements of $\mathcal{E}_i(g_i)$. Thus, $e^1(G^t)$ is guaranteed to be generated. Since the heuristic function is accurate, it will also be chosen for execution.

2. Assume that the claim holds for $e^j(G^t) \mid j = 1, \ldots, n-1$. We have to prove that $e^n(G^t)$ will be generated. According to the induction assumption, $e^{n-1}(G^t)$ will be reached at step $t + n - 1$. Since $e^n(G^t) \models G$ and belongs to $F_{\text{ollow}}^m(e^{n-1}(G^t))$, it must be the case that each g_i^t belongs to $F_{\text{ollow}}^m(e^{n-1}(G^t))$. Thus, each a_i would declare g_i^t at step r, and $e(G^t)$ will be found. □

4.4 Efficiency of the Process

The procedure has the following advantages: **(a)** alternatives are generated by the entire group dynamically (allowing the procedure to be distributed [6]); **(b)** the heuristic function will be calculated for "feasible" alternatives (infeasible alternatives need not be considered, reducing the procedure's computational complexity [7]); **(c)** conflicts and "positive" interactions are addressed within a unified framework.

The process also significantly reduces the complexity of the planning process in comparison to a central planner. The complexity of the planning process is measured by the time (and space) consumed. Let b be the branching factor of the planning problem (the average number of new states that can be generated from a given state by applying a single operator), and let d denote the depth of the problem (the optimal path from the initial state to the goal state). The time complexity of the planning problem is then $O(b^d)$ [11].

In a multi-agent environment, where each agent is capable of carrying out each of the possible operators (possibly with differing costs), the complexity may be even worse. A centralised planner would need to consider assigning each operator to each agent. Thus, finding an optimal plan becomes $O(n \times b)^d$ (where n is the number of agents).

However, since in our scenario the global goal is decomposed into n tasks ($\{g_1, \ldots, g_n\}$) the time complexity is reduced significantly. Let b_i and d_i denote respectively the branching factor and depth of the optimal plan that achieves g_i (in general $b_i \approx \frac{b}{n}$ and $d_i \approx \frac{d}{n}$). Then, as shown by Korf [11], if the tasks are independent or serialisable,[8] the central multi-agent planning time complexity can be reduced to $\sum_i ((n \times b_i)^{d_i})$. And since agents plan in parallel, planning time is further reduced to $\max_i (n \times b_i)^{d_i}$. Moreover, since each agent plans with respect to its own view (actual assignment of operators is done during the merging process itself) the complexity of "sub-planning" becomes $\max_i (b_i)^{d_i}$.

The complexity of the merging process itself is significantly reduced because the branching factor of the search space is strictly constrained by the individual plans' constraints. Second, the Hill-Climbing algorithm is using a relatively good heuristic function, because it is derived "bottom-up" from the plans that the agents have already generated (not simply an artificial h function). Third, generation of successors in each step is split up among the agents (each doing part of the search for a successor).

The algorithm also allows relatively flexible control since changes in the global goal may affect only few members of the multi-agent environment (only they need to replan).

5 Non-Benevolent Agents—Vickrey's Mechanism

Throughout this paper we have assumed that the agents are benevolent. If agents are self-interested (utility maximisers) they should be motivated to contribute to the global plan. In such a case, agents should be paid for their work. Thus, each agent should be paid for performing an operator according to the bid it gives for that operator.

In constructing an optimal plan, it is critical that agents, at each step, express their true cost values (bids). However, if our group consists of autonomous, self-motivated agents, each concerned with its own utility (and not the group's

---

[8] A set of tasks is said to be *independent* if the plans that achieve them do not interact. If the subgoals are *serialisable* then there exists an ordering among them such that achieving any subgoal in the series does not violate any of its preceding subgoals.

welfare), they might be tempted to express false cost values, in an attempt to gain higher payment. This is a classic problem in voting theory: the expression of cost values at each step can be seen as an (iterative) cardinal bidding procedure, and we are interested in a non-manipulable bidding scheme so that the agents will be kept honest. In fact, we want the agents to be honest both in their declaration of constraints during the merging process, and also while bidding during that process.

Fortunately, there do exist solutions to this problem, such that the above plan choice mechanism can be used even when the agents are not necessarily benevolent and honest. By using a variant of the Vickrey Mechanism [17] it is possible to ensure that all agents will bid honestly and will declare their constraints honestly. This is done by minor changes to Step 5 of the procedure given in Section 4.2. As before, an operator would be assigned to the agent that declares the minimal cost (bid), however the actual payment would be equal to the next lowest bid. Note, that under this scheme total payment for the plan would be greater than its actual cost.

To see why declaring the true cost is the dominant strategy, consider a_i facing a bid over Op_k. Assume that the real cost of performing the operator is $X = c_i(Op_k)$. To "win" the operator, a_i might be tempted to bid $X - \delta$, counting on the second-best bid to be $\geq X$. However, it might be the case that the second-best bid is $X - \epsilon$ (such that $\epsilon < \delta$) and a_i would win the bid but lose ϵ. (If the second best bid is $\geq X$ than bidding X would be as good as bidding $X - \delta$, so there is no reason to take the risk.) On the other hand, a_i might be tempted to bid $X + \delta$ in order to be paid more. But doing so, he risks losing the bid altogether; another agent may win with a bid of $X + \epsilon$ and a_i would not be paid at all (while by bidding honestly he would be paid ϵ). If the winning bid will be $X + \delta$, then a_i would gain no advantage since his payment will be determined by the second-best bid (and he might as well bid X).

A formal analysis of the process supports the following claim:

Theorem 2. *At any step t of the procedure, i's best strategy is to bid over the induced actions at that step (all operators in $P(Ex^*(A^{t+}))$) according to his true cost.*

6 Related Work

An approach similar to our own is taken in [13] to find an optimal plan. It is shown there how planning for multiple goals can be done by first generating several plans for each subgoal and then merging these plans. The basic idea there is to try and make a global plan by repeatedly merging plans that achieve the separate subgoals. Finding the solution is guaranteed (under several restrictions) only if a sufficient number of alternative plans are generated for each subgoal. Our approach does away with the need for several plans by treating constraints instead of sequences of actions. We also have agents do the merging in parallel.

In [9] it is shown how to handle positive interactions efficiently among different parts a given plan. The merging process looks for redundant operators (as opposed to aggregating constraints) within the same *grounded linear plan* in a dynamic fashion. To achieve an optimal final plan it takes that algorithm $O(\prod_{i=1}^{n} l(P(g_i)))$ steps, while the approximation algorithm that is presented there takes polynomial time. In [20], on the other hand, it is shown how to handle conflicts efficiently among different parts of a given plan. Conflicts are resolved by transforming the planning search space into a constraint satisfaction problem. The transformation and resolution of conflicts is done using a backtracking algorithm that takes cubic time.

In our framework, both positive and negative interactions are addressed simultaneously. Positive interactions are detected during each aggregation step of the merging process. The most efficient merge is determined by the heuristic value that each extension scores. Similarly, methods for conflict resolution are employed within a single aggregation step: promotion of clobberer, demotion of clobberer, and separation [1] are all considered when a possible extension is generated. Among these possibilities, the optimal one (according to the heuristic function) is chosen. Introduction of a white knight, if necessary, is detected at the next step of aggregation. This phenomenon is enabled by the strong assumption that sub-plans are represented as abstract sets of necessary constraints and that each aggregated extension has an associated realistic heuristic value. The fact that aggregation is done linearly does away with the need for the complete lookahead that is assumed by others' methods. This fact is essential to interleaving the process of planning with execution.

Our approach also resembles the GEMPLAN planning system [12]. There, the search space is divided into "regions" of activity. Planning in each region is done separately, but an important part of the planning process within a region is the updating of its overlapping regions. The example given there, where regions are determined with respect to different expert agents that share the work of achieving a global goal, can also be addressed in a natural way by the algorithm presented here (although we assume that all the domain's resources can be used by the participating agents). This model served as a basis for the DCONSA system [14] where agents were not assumed to have complete information about their local environments. The combination of local plans was done through "interaction constraints" that were pre-specified.

7 Conclusions

In this paper, we presented a novel multi-agent interleaved planning process. The process relies on a dynamic distribution of individual tasks. Agents solve local tasks, and then merge them into a global plan. By making use of the computational power of multiple agents working in parallel, the process is able to reduce the total elapsed time for planning as compared to a central planner. The optimality of the procedure is dependent on several heuristic aspects, but in general increased effort on the part of the planners can result in superior global plans.

The techniques presented here are also applicable in any situation where multiple agents are attempting to merge their individual plans or resolve inter-agent conflict when execution time is critical.

Acknowledgments

This work has been partially supported by the Air Force Office of Scientific Research (Contract F49620-92-J-0422), by the Rome Laboratory (RL) of the Air Force Material Command and the Defense Advanced Research Projects Agency (Contract F30602-93-C-0038), by an NSF Young Investigator's Award (IRI-9258392) to Prof. Martha Pollack, and by the and by the Israel academy of sciences and humanities (Wolfson Grant), and by the Israeli Ministry of Science and Technology (Grant 032-8284).

References

1. D. Chapman. Planning for conjunctive goals. *Artificial Intelligence*, 32(3):333–377, July 1987.
2. D. Corkill. Hierarchical planning in a distributed environment. In *Proceedings of the Sixth International Joint Conference on Artificial Intelligence*, pages 168–175, Tokyo, August 1979.
3. E. H. Durfee and V. R. Lesser. Using partial global plans to coordinate distributed problem solvers. In *Proceedings of the Tenth International Joint Conference on Artificial Intelligence*, pages 875–883, Milan, 1987.
4. E. H. Durfee and V. R. Lesser. Negotiating task decomposition and allocation using partial global planning. In Les Gasser and Michael N. Huhns, editors, *Distributed Artificial Intelligence, Vol. II*, pages 229–243. Morgan Kaufmann, San Mateo, California, 1989.
5. Edmund H. Durfee. *Coordination of Distributed Problem Solvers*. Kluwer Academic Publishers, Boston, 1988.
6. E. Ephrati and J. S. Rosenschein. Distributed Consensus Mechanisms for Self-Interested Heterogeneous Agents. In *First International Conference on Intelligent and Cooperative Information Systems*, pages 71–79, Rotterdam, The Netherlands, May 1993.
7. E. Ephrati and J. S. Rosenschein. Reaching agreement through partial revelation of preferences. In *Proceedings of the Tenth European Conference on Artificial Intelligence*, pages 229–233, Vienna, Austria, August 1992.
8. J. J. Finger. *Exploiting Constraints in Design Synthesis*. PhD thesis, Stanford University, Stanford, CA, 1986.
9. D. E. Foulser, M. Li, and Q. Yang. Theory and algorithms for plan merging. *Artificial Intelligence*, 57:143–181, 1992.
10. Matthew J. Katz and J. S. Rosenschein. Verifying plans for multiple agents. *Journal of Experimental and Theoretical Artificial Intelligence*, 5:39–56, 1993.
11. R. E. Korf. Planning as search: A quantitative approach. *Artificial Intelligence*, 33:65–88, 1987.
12. A. L. Lansky. Localized search for controlling automated reasoning. In *Proceedings of the Workshop on Innovative Approaches to Planning, Scheduling and Control*, pages 115–125, San Diego, California, November 1990.

13. D. S. Nau, Q. Yang, and J. Hendler. Optimization of multiple-goal plans with limited interaction. In *Proceedings of the Workshop on Innovative Approaches to Planning, Scheduling and Control*, pages 160–165, San Diego, California, November 1990.

14. R. P. Pope, S. E. Conry, and R. A. Mayer. Distributing the planning process in a dynamic environment. In *Proceedings of the Eleventh International Workshop on Distributed Artificial Intelligence*, pages 317–331, Glen Arbor, Michigan, February 1992.

15. Reid G. Smith. *A Framework for Problem Solving in a Distributed Processing Environment*. PhD thesis, Stanford University, 1978.

16. G. J. Sussman. *A Computational Model of Skill Acquisition*. American Elsevier, New York, 1975.

17. W. Vickrey. Counterspeculation, auctions and competitive sealed tenders. *Journal of Finance*, 16:8–37, 1961.

18. Frank von Martial. Multiagent plan relationships. In *Proceedings of the Ninth International Workshop on Distributed Artificial Intelligence*, pages 59–72, Rosario Resort, Eastsound, Washington, September 1989.

19. Frank von Martial. Coordination of plans in multiagent worlds by taking advantage of the favor relation. In *Proceedings of the Tenth International Workshop on Distributed Artificial Intelligence*, Bandera, Texas, October 1990.

20. Q. Yang. A theory of conflict resolution in planning. *Artificial Intelligence*, 58(1-3):361–393, December 1992.

Multi-agent communication

Generic, Configurable, Cooperation Protocols for Multi-Agent Systems

Birgit Burmeister   Afsaneh Haddadi   Kurt Sundermeyer

Daimler-Benz AG
Research and Technology
D-10559 Berlin
Alt-Moabit 91b
Phone: +49 30 399 82 – 202 / –236
Email: [bur I afsaneh I sun]@DBresearch–berlin.de

Abstract

In this paper we propose a unified and general mechanism for developing cooperation protocols in multi-agent systems. The protocols are essentially speech act based but have considerable advantages as compared to previous approaches: First, they are generic in the sense that a protocol execution algorithm can treat the domain independent parts separately from the application dependent reasoning and deciding processes involved. And second, they are recursively defined from primitives which allow a designer (or eventually the agents themselves) configure the appropriate general or domain-specific cooperation protocols.

1   Introduction

In the DAI community it is widely agreed that cooperation among reasoning agents needs explicit communication. It is also agreed that communication in DAI is more than communication in traditional distributed systems. But whereas traditional communication already is standardized, there is not yet a common agreement about how communication should be treated in DAI. As a matter of fact a literature survey is rather disappointing in that many authors have been concerned with task-specific protocols. Furthermore protocols by different authors are incomparable, because each of them interrelates low level aspects (e.g. syntax of messages) and the aspects of a cooperation dialogue in a different way.

If communication would solely be described in terms of sending and receiving messages, each agent must be able to infer what the sender intended by uttering a message (e.g. whether it is a query about some information or a command to act). If messages are not structured, making this inference could be very inefficient. Thus messages should be bound by formal restrictions and structured for the ease of interpretation, for instance by employing message types, such that the intention of the sender could be immediately recognized from the message itself. This alone would still not be sufficient, since agents also ought to know how to react to a message, or what to expect after sending a message. All these requirements should be met collectively by a unique framework. We call such a framework for structuring dialogues among agents "cooperation protocols", although they do not only apply to truly benevolent cooperative situations, but also to situations with elements of competitiveness.

In the work described in this article we aim at devising a method in order to rapidly develop, prototype and experiment with various cooperation protocols. To do this one should first be able to answer the questions: (1) how rich the syntax of a message should be to aid its "fast" interpretation? (2) is it possible to define application independent protocols? and (3) could one find a set of primitives and build protocols from these primitives?

In view of the wide span of rather complex cooperation patterns (like "contracting") to primitive communication forms (like "inform") it seems convenient to distinguish various layers, namely

– structure/syntax of messages;
– message types (like inform, query, demand, ...);
– procedures for preparing messages to send and for processing received messages;
– protocols as frameworks for dialogue in a cooperation;
– cooperation patterns (like contracting, bargaining, persuading, resolving conflicts);
– mechanisms to select and keep track of protocols and cooperation patterns.

In this article we shall concentrate on the first four layers. We have developed a syntax and semantics for messages in terms of message types which will be described in Section 3. This section also includes the third layer, namely procedures for sending and processing messages. The fourth layer concerning protocols and dialogues structured by protocols, involves devising an appropriate representation of protocols and how they could be made operational. Hence Section 4 includes details of protocol representation, an algorithm for executing protocols, and a description of how the protocols can be designed within DASEDIS, our simulation and development environment for multi-agent systems. Section 5 contains a comparison of our approach to related works. The paper is concluded with further directions to this work in Section 6. But first we give a review of previous approaches to cooperation and communication, in order to set the background and to clarify what we would like to attempt.

2   From Speech Acts to Protocols

A literature survey in "cooperation and communication" reveals that the majority of approaches have relied on speech act theory. Due to [Austin 62], is the distinction of *locutionary, illocutionary* and *perlocutionary* aspects in a speech act. By this, the simple utterance of a sentence (locutionary) is distinguished from its intended effects (illocutionary) and its actual effects (perlocutionary) on the receiver. Inevitably the actual securing of the perlocutionary effects is beyond the control of the speaker. The locutionary aspects of a speech act falls within the realm of standard communication theory, where it is well studied. Of interest therefore, in the communication in multi-agent systems, are the illocutionary acts.

[Cohen/Levesque 90] argue that properties of illocutionary acts can be derived from the speaker's and hearer's mental states. Although this has lead to some interesting results, we believe that since artificial agents have the benefit of using artificial languages, the illocutionary aspects can be explicitly coded in the message in form of message types.

The classification of illocutionary acts into *assertives, directives, commissives, declaratives* and *expressives* by [Searle 69] has given the inspiration to the researchers in DAI for the choice and nature of message types. One of the early advocates were [Cohen/Perault 79] who used the message types "inform" and "request" to model assertives and directives in a plan calculus for communication. Since then many message types have been introduced and employed, but apart from the KQML proposal [Shapiro/Chalupsky 92], no attempt has been made on the standardization of message types. The aim of this first KQML proposal was to standardize the communication of knowledge bases and thus is restricted to communication involved with exchanging knowledge. In our opinion, the proposal is not sufficient for multi-agent systems, where there also is communication about actions, intentions, and resources other than knowledge.

Searle's classification received criticism from various sides (see e.g. [Levinson 81]). The distinctions among different classes are not made on a principled basis, and the classification is doubtful in that some communication forms fall into different classes. As far as multi-agent systems are concerned, there is no need for declaratives and expressives and the remaining classes are still too rough for structuring messages. Finally and most importantly, speech acts alone cannot represent a dialogue.

[Ballmer/Brennenstuhl 81] faced enormous difficulties when attempting to classify all German speech act verbs into Searle's classes. Instead they suggested to group the verbs that are similar in meaning, into semantic categories, and group the semantic categories into models. For example a model of an "enquiry", consists of a "question" and an "answer". The models also include an ordering of the categories according to their temporal relationship and degree of strength. The categories and the models lead to dialogues and appropriate frameworks for structuring dialogues ("protocols").

Number of researchers have proposed such frameworks, like the well known Contract Net Protocol [Smith 80] and the hierarchical protocol [Durfee/Montgomery 90]. These frameworks, however, have been mostly concerned with specific aspects of cooperation. For instance the Contract Net Protocol is designed for task-allocation, and the hierarchical protocol aims at coordination.

Some authors have devised general specification languages to describe cooperation independent of the implementation language, e.g. [deGreef et al. 92]. The purpose of their language is to abstract away from details which are not relevant to the interacting agents. If this language is as claimed general enough for specification of all types of "cooperation methods", its merits are doubtless. Nevertheless, of benefit would be some means by which cooperation methods (i.e. what we call cooperation protocols) can be conveniently developed from their components and consequently analyzed and experimented with at a more abstract level.

In summary, the studies of communications between agents owe a great deal to the work in speech act theory, specifically, their impact on the message types used in communication in multi-agent systems. Numerous message types have been introduced in the literature, however, each author has his/her own interpretation of a message type. To study cooperation, one requires protocols which go beyond the details of communication, leading to dialogues and cooperation protocols. Individual protocols have been devised for specific tasks. Inevitably the nature of cooperation differs from one application to another. To best study cooperation protocols and experiment with them, one requires general means to: (1) represent and sketch the phases of a cooperation protocol (2) develop these protocols from previously defined components in a modular manner, and (3) make them operational. The basis would be a general specification language to describe cooperation protocols independent of the implementation language. This language should be further complemented with graphical representations to provide more convenient tools for developing and analyzing cooperation protocols.

3 Messages

3.1 Syntax and Semantics of Messages

The basic elements of communication are messages. For ease of message interpretation in multi-agent communication, we have introduced a fixed syntax for messages. A message consists of a *header* and a *content*. The header contains administrative information such as addresses of the sender and the receiver. The semantics of a message is described by the *content* which includes *message type, descriptor, text* and *agent* .

The illocutionary force of messages is granted by message types. Due to the lack of appropriate classifications of message types, we have chosen those message types which we believe are essential in the context of most cooperation protocols. These are INFORM, QUERY and COMMAND, which are similar in meaning with those message types used by other authors, e.g. provide-information, request-information, request-action [Doran 85], inform, question, request [Huhns et al. 90] and inform, query, request [Numaoka/Tokoro 90].

As [Feierstein 92] illustrates, message types can be combined with what he terms *generic objects* to further assist the message interpretation, for instance, "INFORM goal" or "COMMAND task". These objects in our terms are called *descriptors* which describe what the proposition of the message is about. Since our agent concept is based on the triple (*resources, behavior, intentions*) [Sundermeyer 90] messages are generally either about resources, behavior (or plan of actions) or intentions (such as long-term goals, attitudes, roles etc.). Descriptor is therefore one of resource, behavior or intention. The other two sub-fields of content, namely *text* and *agent* are respectively, an application dependent syntactic construct for the actual proposition, and the agent which the proposition refers to. The *agent* can be the sender, receiver, or another agent.

For instance a message consisting of message type INFORM, descriptor "behavior" and text "(move :from 'table :to 'chair)", informs the receiver that the sender is currently moving from the table to the chair. Or a message with type QUERY, descriptor "resource" and text "color" means that the sender needs to know the color of the receiver.

3.2 Sending and Processing Messages

The next stage concerned with communication is sending and processing messages.

Messages are prepared and sent by *send-procedures*. Each procedure is specialized to prepare messages with a specific message type and instantiate the required fields of the message.

Similarly for each message type there is a *process-procedure*. A process-procedure's task is to interpret and handle responds to received messages. These procedures however handle general default responds, e.g. transfer the content of an INFORM-message into the knowledge base or retrieve the answer to a QUERY from the knowledge base. Any further, application specific responds will have to be specified by the designer.

4 Protocols

4.1 Representation of Protocols

As was mentioned in the introduction, we attempt to arrive at a unified approach for defining and conveniently developing cooperation protocols. As a first step, this requires an expressive representation for sketching dialogues. More essentially, the representation should be also immediately translatable to the implementable equivalent. Next, the sender and receiver of a message, initiating a form of cooperation, should be able to use the same protocol. This seems useful as the relation of the actions taken by the dialogue participants during a particular course of interaction is explicitly represented and because it allows for a more compact representation of the dialogue. In this respect, the representation should also take into account that both participants of a dialogue will use a copy of the same protocol.

Fig. 1 shows the graphical representation of a generic protocol. The representation of a protocol consists of a heading which is the name of the protocol and some parameters, and a labeled tree which represents the possible steps of dialogue in that protocol.

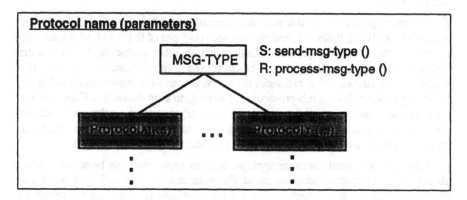

Fig. 1. A Generic Protocol

The name of the protocol reflects the nature of the protocol. Therefore the names of the primitive protocols presented later, are the names of the message type used in the very first message sent in the protocol, and an 'ing' ending (to distinguish it from a message type). For example an Informing protocol's initiating message is INFORM and Querying protocol's is QUERY.

The parameters of the protocol are the addresses of the sender and receiver of the first message in the protocol, and two parameter lists. One list is employed by the initial sender (e.g. the content of the message) and the other by the initial receiver (e.g. the message received) in a protocol. Due to lack of space parameters of protocols and the procedures mentioned have been omitted in the remaining of this paper.

The nodes in the tree represent dialogue states, links represent transitions from a dialogue state to another, and branches represent alternative transitions.

For ease of readability and consistency, nodes in a protocol tree are designated as follows:

- The root node characterizes the protocol, it is labeled by the message-type which is unique for that protocol. For every protocol there has to be a unique message type to distinguish it from another protocol. The root node represents the state when the first message is sent by the sender and processed after being received by the receiver. This is represented by denoting the send-procedure after the "S:"-label and the corresponding process-procedure after the "R:"-label.

- Other nodes are calls to other protocols (sub-protocols). The texts on these nodes are (sub-)protocol names with their parameters.

- The color of a node represents the active agent (or sender of the message) at that node. A white node indicates that the sender of the first message in the protocol is also the sender at this node. A shaded node indicates that the receiver of the first message in the protocol is the sender at this node.

In Fig. 3, three primitive protocols are presented. These protocols can be used as building blocks in more complex protocols, and for simple interactions concerning enquiring or giving information, and commanding an action. These are the protocols dealing with the basic message types INFORM, QUERY, and COMMAND introduced in section 3.1.

The Informing protocol is used to provide information to the receiver. Messages of type ANSWER, REJECT, REPORT, and the like, are sub-types of INFORM, which are also treated by the Informing protocol when called in other protocols, to send some specialized information. A Querying protocol is used to enquire some information. The receiver of the first message, processes the query, prepares the answer and invokes the Informing protocol (as the sub-protocol of Querying) to send the answer. Commanding is used to command an agent (by default the receiver) to execute a behavior. When the other agent executes the behavior it sends a report to the commander. The report can range from state of the execution, resources that become sparse and so forth.

If required, other primitives more appropriate to an application can be defined, or an already defined primitive can be modified. For instance in a Commanding protocol, one can include appropriate responds after the report is sent or eliminate the option of rejecting the command.

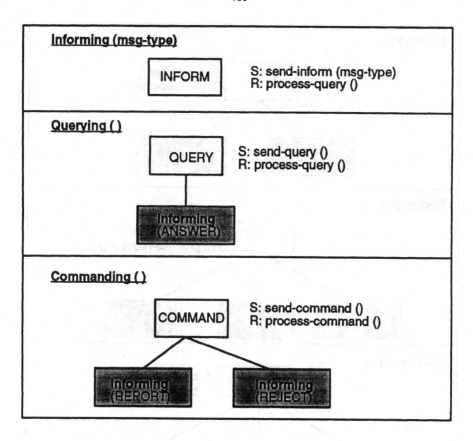

Fig. 3: Primitive Protocols

In addition these primitive protocols are used to construct more complex cooperation protocols. To demonstrate how this is done, Fig. 4 represents three protocols Offering, Requesting and Proposing.

An Offering protocol is used when an agent offers to execute a behavior for another agent. The possible reactions of the receiver of the offer are (1) accept the offer by commanding the execution of the offered behavior; or (2) reject the offer. This protocol is the counterpart of the Commanding protocol.

The Requesting and Proposing protocols are similar to Commanding and Offering but allow for negotiation. The Requesting protocol is used by an agent to request another agent to execute a behavior, the possible reactions of the receiver of the request are (1) reject the request; (2) accept the request by using an Offering protocol which in turn gives the receiver of the offer (i.e., the sender of the request), options of rejecting the offer (i.e., withdrawing the request), or commanding it; or (3) propose the requested action with modified conditions to the requester.

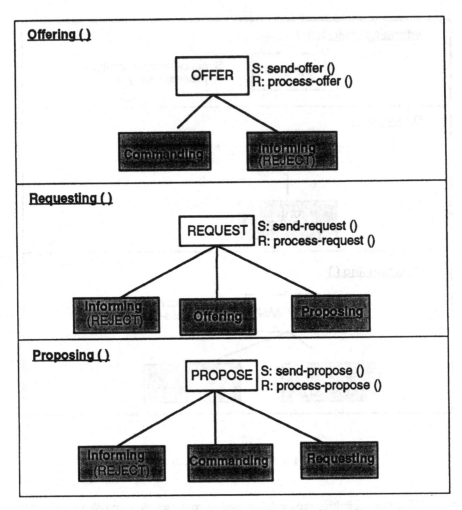

Fig. 4: Offering, Requesting, and Proposing Protocols

In a Proposing protocol, used by an agent to propose the execution of a behavior for another agent, the reactions of the receiver of the proposal are (1) reject the proposal; (2) accept the proposal by using a Commanding protocol which in turn gives the receiver of the command options of rejecting the command (i.e., withdrawing the proposal), or simply carrying out the action that was proposed; or (3) send a modified proposal, stating for example that the proposed action should be carried out under different conditions.

In [Haddadi/Sundermeyer 93] we classified the situations in which an agent initiates an interaction with another agent into support and hindrance situations. That is, an agent requires support of other agents, it wants to support some other agent or it is hindered by other agents from performing its task. In analyzing these situations, we found that the aforementioned protocols are in general appropriate to these situations. For instance, when an agent wants support of another agent the Commanding or Requesting protocol can be used. In case an agent wants to offer its support to another agent, the Offering or Proposing protocol is appropriate. In hindrance situations the Requesting or Proposing protocol can be used to negotiate with the hindering agent to avoid the hindrance.

Although we think that the protocols described above are general enough to be used in a large number of interaction situations, our approach allows for the construction of (more complex) application or task specific protocols.

As an example we show the well-known Contract Net Protocol. The protocol used by the manager and a possible bidder in a Contract Net may be defined as in Fig. 5. The first left branch denotes that the receiver of the announcement would not participate in bidding. Another way of defining this protocol would be to simplify the tree by using an Offering protocol for the bidding sub-tree.

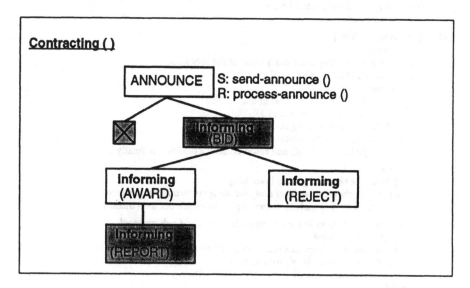

Fig. 5: Contract Net Protocol

4.2 Protocol Execution Algorithm

The protocol execution algorithm (PE) described in this section makes the protocol components operational, i.e. it prepares the appropriate parameters and calls the concerned send- or process-procedures and sub-protocols denoted at the different nodes of a protocol. Any decision making necessary in the course of a protocol execution (i.e. at the branching points of the tree), is handed over to a specific decision making component and is not part of the algorithm.

Cooperation protocols can be thought of as plans of actions, and the PE as a plan executer. Hence at the sending side, the PE is activated when the sender initiates a dialogue for cooperation, and at the receiving side it is activated when a message is received. The pseudo-code representation of the algorithm is given in Fig. 6. In this article the algorithm is simplified for clarity, by assuming that the role of the active agent changes from one level to the next, although the representation is not restricted in this way.

Protocol-Execution (protocol, mode, :open)
    { *protocol*: the protocol to execute,
        mode: sending or receiving mode
        :open: the protocol is already open}

begin

– if mode = :send then
    – call the sending procedure designated at the root node,
    – if this is a leaf node of protocol then
        – close protocol

– else { mode is :receive }
    – if not :open then
        – call the process procedure designated at the root node
        – if this is a leaf node of protocol then
            – close protocol
        – else   { there is a next level in the
                            graph of the protocol }
            – If there is a branching then
                –pass control to decision component
            { PE will be suspended, and resumed after a decision is finally made }

            – endif
            { Whether branching or no branching }
            – traverse the tree one level and find the appropriate sub-protocol
                { In case of branching traversing is led by the decision made }

            – store parameters in the appropriate slots of the sub-protocol
            – set mode to :send
            – call Protocol-Execution (sub-protocol and mode)
            – if this is a leaf node of protocol then
                – close the protocol
        – endif
    – else {mode = :receive and :open}
        – traverse the tree one level and find the appropriate sub-protocol
        { In case of branching traversing is led by the type of the received message }

        – store parameters in the appropriate slots of sub-protocol
        – set mode to :receive
        – call Protocol-Execution (sub-protocol, mode)
        – if this is a leaf node of protocol then
            – close protocol
    – endif
– endif

end.

Fig. 6: Protocol Execution Algorithm

The algorithm reflects the two possible modes an agent can be in within a protocol, namely sending or receiving messages. When a protocol is executed in the sending mode, the send-procedure at the root node is called and the protocol is suspended until a message referring to this protocol is received. When a message is received and it refers to an already activated protocol, the tree of that protocol is traversed one level. Otherwise, first the processing-procedure at the root node is called and then after a decision is made (outside of the protocol), the protocol is traversed to the next level. The simplicity of the algorithm is due to the generality and expressiveness of the protocol representation. The algorithm can be used for any arbitrary complex protocols.

4.3 Protocols in the DASEDIS-testbed

The cooperation protocols are building blocks in the testbed DASEDIS, which has been developed in our ongoing project COSY [Burmeister/Sundermeyer 90]. Underlying this testbed is a modular agent architecture, which reflects the sensoric, actoric, communicative, intentional and cognitive features of an agent. The cognitive capabilities are represented in a module COGNITION which is realized as a knowledge-based system. The protocol execution algorithm is an enhancement of the cooperative problem solving part of COGNITION, as described in [Burmeister/Sundermeyer 92]. The protocols known to an agent are part of its knowledge-base.

Within DASEDIS, the designer of a multi-agent-system can use the predefined cooperation protocols or can design his/her own protocols. Designing a protocol is done in three steps: (1) define the protocol by its graphical representation, (2) define any new message-types and (3) define the send- and process-procedures.

The designers are provided means by which they can convert the graphical representation of a protocol to the equivalent LISP-representation. Message type(s) associated with the protocol are defined as subtypes of a general message type or of previously defined message types. For each message type there is a send- and a process-procedure, which the designer has to define. When a message type is a subtype of another message type, by default, the sending and processing procedures of the super-type will be used. The designer can however develop specialized processing procedures.

The testbed DASEDIS contains tools for developing agents and multi-agent scenarios. In addition to the already available tools we will devise a graphical editor which automatically translates the cooperation protocol diagrams to a LISP-representation, and furthermore simplifies the steps (2) and (3) mentioned above. In this way the configuration of new protocols from already existing ones can be achieved more easily.

5 Related Work

Various authors make use of graphical representations for protocols. These are basically state transition diagrams, but according to the aspect of cooperation in interest, the meaning of the graphs differ from one author to another. In general the nodes represent the state of conversation and the arcs the messages exchanged. [Winograd/Flores 86] represent and analyse dialogues by networks of speech acts. As far as representation is concerned their graphs are very expressive and easy to follow. However, they still miss the information required to make them operational. [Kreifelts/von Martial 91] started with similar diagrams and subsequently replaced them by a Petri-net, in order to solve the problem of "overtaking" and "crossing" messages. Unfortunately the expressive power of the original diagrams is lost and the protocol components and steps are no longer visible. The new diagram is rather a general diagram stating the communication between two agents. We see the problems considered by Kreifelts and Martial as being related to the level of physical communication, and thus not appropriate at the level of protocol representation. [Campbell/D'Inverno 90] proposed a general graphical representation in which the protocols not only show the communication between two agents, but also the reasoning involved in the course of a protocol. For this purpose, nodes represent the reasoning states. Although these graphs may well serve their purpose, the concept of adopting such a notation as a general means of representing protocols is doubtful, since it is questionable if reasoning behind each node is purely application-independent. The closest to our approach is the notation used by [Chang/Woo 92] in SANP (Speech Act Negotiation Protocol). The similarity lies on the fact that they employ modular representation, by allowing nodes in the diagram to be either single states of conversation, or 'calls' to other protocols (or what they term "phases").

In summary, our protocol graphs represent the possible conversation courses, hiding away the reasoning states, as opposed to Campbell and D'Inverno's diagrams. We believe that for generic protocols, since some of the reasoning and decision processes involved in the course of a cooperation are application dependent, they should not be present at the representation level. Our protocols can be seen as an abstraction from those state transition diagrams where a node denotes either the sender or the receiver of a message. In our graphs, the sending, receiving and processing of a message is adhered to a single node and the transition to the next state is made when sending or receiving a new message. In the same lines as Chang and Woo, we adopt a modular representation where nodes can also be calls to other protocols. The modularity and the abstraction allows rapid prototyping of larger protocols.

The comparison with other works can only be made at representational level of dialogues and protocols. Since we aim at devising tools to conveniently develop and experiment with various cooperation frameworks, we were not only concerned with the expressive power of graphs at representational level, but also at operational level. Therefore our graphs are not only demonstrative for sketching cooperation protocols for analysis, but are easily translatable to an executable representation.

6 Conclusion and Outlook

We described a global control structure for cooperation and communication in terms of cooperation protocols. These protocols are built from message types and procedures for sending and processing messages. In this paper we have

- defined messages as structured by message types and descriptors (intentions, resources, behavior)
- separated the application specific aspects of cooperation and communication from the domain-independent pieces in the protocols
- shown how cooperation protocols can be simply developed from primitive concepts.

Since the cooperation protocols we proposed have a clear syntax and operational semantics, they can be seen as a proposal for a standard for dialogues in DAI. As long as a systems designer decides for dealing with the reasoning and decision processes outside the protocol, he/she can use the representation and execution algorithm of the cooperation protocols. This is independent of which message types are used, as long as the semantics accords with the semantics underlying our approach. This implies that our approach can easily adopt any standardization of message types the DAI community might agree upon.

Whilst we were able to define generic configurable cooperation protocols some pieces are still missing:

- Protocols are essentially bound to message types and not immediately to cooperation patterns like contracting, bargaining, persuading. Each of these can in principle be treated by different strategies. There has to be a methodology for associating protocols for each strategy.
- There are to be criteria for protocol selection as well as intervening from one protocol to another. An agent should also keep a history of its dialogue while interacting with an agent in order to avoid unnecessary repetitions in communication.

As we are interested in autonomous agents, the ultimate goal of the work in future will be concerned with active protocol selection, that is, an agent selects the protocol most appropriate to its intentions and to its knowledge about the surrounding. The foundation for this kind of autonomy was laid in [Haddadi/Sundermeyer 93] in terms of intentional and epistemic states of an agent.

In future our cooperation protocols could serve to specify the communication in agent-oriented programming, as introduced by [Shoham 93], namely enhancing the simple message-passing between mentally inert objects in object-oriented programming, by communication types for agents with mental states.

Acknowledgements

We would like to thank the other members of the COSY team, especially Olaf Schreck for many inspiring ideas concerning the protocol graphs.

References

[Austin 62] J.L.Austin, How to do things with words, Oxford Univ. Press, 1962

[Ballmer/Brennenstuhl 81] T. Ballmer, W. Brennenstuhl, Speech Act Classification: A Study in the Lexical Analysis of English Speech Activity Verbs, Springer, Berlin, Heidelberg, 1981

[Burmeister/Sundermeyer 90] B.Burmeister, K.Sundermeyer, "COSY: Towards a Methodology of Multi-Agent Systems", Draft Proc. CKBS-90, 1990

[Burmeister/Sundermeyer 92] B.Burmeister, K.Sundermeyer, "Cooperative Problem-Solving Guided by Intentions and Perception" in [Demazeau/Müller 92]

[Campbell/D'Inverno 90] J.A.Campbell, M.P.D'Inverno, "Knowledge Interchange Protocols" in Y.Demazeau, J.P.Müller (eds.), Decentralized A.I., Elsevier/North-Holland, 1990

[Chang/Woo 92] Man Kit Chang, C.C.Woo, "SANP: A Communication Level Protocol for Negotiations" in [Demazeau/Müller 92]

[Cohen/Levesque 90] P.R.Cohen, H.J.Levesque, "Rational Interaction as the Basis for Communication" in P.R.Cohen, J,Morgan, M.E.Pollack (eds.), Intentions in Communication, MIT Press, Cambridge, 1990

[Cohen/Perrault 79] P.R.Cohen, C.R.Perrault, "Elements of a Plan-Based Theory of Speech Acts" Cognitive Science, 3, Ablex Publ.Corp., 1979

[Demazeau/Müller 92] Y.Demazeau, J.P.Müller (eds.), Decentralized A.I. 3, Elsevier/North-Holland, 1992

[Doran 85] J.E.Doran, "The computational approach to knowledge, communication, and structure in multi-actor systems" in G.N.Gilbert, C.Heath (eds.), Artificial Intelligence and Sociology, Gower, London, 1985

[Durfee/Montgomery 90] E.H.Durfee, T.A.Montgomery, "A Hierarchical Protocol for Coordinating Multiagent Behavior", Proc. AAAI-90

[Feierstein 92] M.Feierstein, "A Model of Cooperative Action", IMAGINE Technical Report TR No.6, 1992

[deGreef et al. 92] P. de Greef, K.Clark, F.McCabe, "Toward a Specification Language for Cooperation Methods", Contribution to GWAI-92 Workshop "Supporting Collaborative Work Between Human Experts and Intelligent Cooperative Information Systems", 1992

[Haddadi/Sundermeyer 93] A.Haddadi, K.Sundermeyer, "Knowledge about Other Agents in Heterogeneous Dynamic Domains", Proceedings ICICIS, Rotterdam 1993

[Huhns et al. 90] M.N.Huhns, D.M.Bridgeland, N.V.Arni, "A DAI Communicative Aide" in M.N.Huhns (ed.), "Proc. of the 10th International Workshop on Distributed Artificial Intelligence", MCC Technical Report Number ACT-AI-355-90

[Kreifelts/vonMartial 91] Th.Kreifelts, F. von Martial, "A Negotiation Framework for Autonomous Agents" in Y.Demazeau, J.P.Müller (eds.), Decentralized A.I. 2, Elsevier/North-Holland, 1991

[Levinson 81] S.C.Levinson, "The Essential Inadequacies of Speech Act Models of Dialogue", in H.Parrett (ed.), Possibilities and Limitations of Pragmatics, J.Benjamin, Amsterdam, 1981

[Numaoka/Tokoro 90], C.Numaoka, M.Tokoro, "Conversation among Situated Agents" in M.N.Huhns (ed.), "Proc. of the 10th International Workshop on Distributed Artificial · Intelligence", MCC Technical Report Number ACT-AI-355-90

[Searle 69] J.R.Searle, Speech Acts, Cambridge Univ.Press, 1969

[Shapiro/Chalupsky 92] S.C. Shapiro, H. Chalupsky, "KQML – Issues and Review", Preliminary Report, State University of New York, Buffalo, 1992

[Shoham 93] Y.Shoham, "Agent Oriented Programming", Artificial Intelligence (60), Elsevier, 1993

[Smith 80] R.G.Smith, "The Contract Net Protocol: High-Level Communication and Control in a Distributed Problem Solver", IEEE Transactions on Computers, C-29, 1980

[Sundermeyer 90] K.Sundermeyer, "Modellierung von Szenarien kooperierender Akteure" in H.Marburger (ed.), GWAI-90, Springer, Informatik Fachberichte, 1990

[Winograd/Flores 86] T.Winograd, F. Flores, Understanding Computers and Cognition, Addison-Wesley, 1986

Around the Architectural Agent Approach to Model Conversations

Milton Corrêa and Helder Coelho

INESC - 9, rua Alves Redol, 7º, 1000 Lisboa, Portugal
email: mcor@inesc.ctt.pt and hcoelho@inesc.ctt.pt

Abstract. Modelling conversations is a rather complex task, constrained by several non-isolated issues and existing within and across different scientific disciplines. In the present paper we propose to cut through this reality, observing how the functional side interacts with the structural side. Our main thesis rests upon the idea that the architecture of the agents is a prerequisite for understanding the structure of intelligence in conversations and that conversations are based upon intentional action. Also, we believe that some mental states (e.g.. belief, desire, intention and expectation) are due to be in the kernel of the agent model, in order to structure its architecture. They strongly influence the interactional capability of any agent. So, any structure of interactions among agents is built by relating the key notions of agent, conversational context and conversational patterns to the mental states of the agents.

These ideas are currently under implementation and validation. A brief outlook of this direction of research and experimentation is shown at the end of this paper.

1 Introduction

The design of autonomous and intelligent agents able to interact in a way close to human beings is not an easy task. However, the motivation is neat, because there is an increasing demand to build societies using advanced information and knowledge systems along distributed settings, either in high technology buildings, electronic offices or in automated factories.

Several lines of inquiry can be set up around the study of dialogues. Often, it is unclear how the communicative actions are related to partner modelling and information processing itself. As a matter of fact, sometimes a syntactical point of view is proposed and there concentration on dialogue patterns with the sole purpose of transferring initiatives (Coelho, 1982; Sernadas et al, 1987). Otherwise, the focus of attention is on the semantic side, and it is concerned with certain factual information (Bunt, 1989). This second research path is very common in the literature (Gaspar, 1990;1991), because those dialogues are popular in real life, such as at information desks in airports and railway stations (Allen, 1990), or with the information service of a telephone company. These two directions can be combined within the set of structural theories.

Our point of view can be classified on the linguistic pragmatic side (Galliers, 1989) and (Cohen and Levesque,1990), alongside the psychological theories, because we propose a close link between the architecture of an agent (structured by mental states) and its dialoguing capability (Sloman, 1991) through speech acts. Also, we introduce the idea of conversational context and we keep active components in parallel (the so-called local agents nested in a "Russian doll" fashion to create the global agent) inside the whole organisation of the agent, each one in charge of satisfying certain rational skills (reasoning, knowledge acquisition and belief revision).

First of all, we consider that communications among autonomous and intelligent agents take on many forms, from the simple word or sentence in a casual chat (exchanges) to the complex net of formal interactions which occur in the course of a conversation between two or more agents through forms and associated procedures. We foresee these communications at large, including them in the most general case of those interactions with the external environment (real world). Secondly, the agent architecture and the definition of conversational context are based upon five basic notions: action, belief (we assume, as an hypothesis, that knowledge and belief are merged in one sole notion), intention, desire, and expectation. The motivation for this is straightforward: we believe that the sequence of communicative actions performed by each artificial agent is related with its mental states (belief, intention, desire and expectation), alongside the theories of rational agents design (Rao and Georgeff, 1991a; 1991b). And, finally, we envisage the conversational context of some interaction as a pattern able to link actions (à la Bateson), i.e. an intention viewed as a temporal sequence of the sub-set of the mental states of the agent and of the sub-set of states of the external world, and a general strategy for executing this sequence according to that intention.

2 A theory around the agent

Behind this effort to model dialogues among autonomous agents (Coelho et al, 1992) there is a strong need to discuss in detail the theoretical and experimental aspects of the required representation, because we may face hard problems when defining the semantic content of the actions and the computational tractability of the overall dialogue processing (Coelho, 1990).

Consider the simple case of knowledge communication among agents (e.g. robots in the ground floor of a factory, bureaucratic artificial agents supporting administrative officers in some organisation), in order to avoid the use of natural language and its consequent linguistic problems. So, we can concentrate on agent modelling dialogue structures, planning efforts and control management within context spaces.

Some authors like (Bunt, 1989) prefer to adopt a strategy to set up a theoretical building, where either communicative actions or dialogue control acts are defined in terms of context changes. Our main idea is quite in the opposite direction. We start by defining the agent architecture taking into account primitive notions and its internal states, then we introduce the notion of conversational context, and, finally, we present the dialogue structures or the organisation of dialogue in terms of mental states. Therefore, agent modelling is a prerequisite for the ability to represent the agent's mental states.

From our observations on experimental interactions (Coelho et al., 1992), we could identify the most appropriate dimensions of the kernel notion of context that are important for the sort of communication that we are interested in. So, the knowledge (or belief) of those agents was selected as the most intuitive notion, either to model the skills of a rational agent (Gaspar, 1990), say perceptual, reasoning, action and communication abilities, or to model its mental states.

Although the set of beliefs can be extended to incorporate the notion of intention (Bunt, 1989), we prefer a much larger tuple, able to include the basic mental states of the agents. Therefore, the communicative force of a dialogue is built up as a mathematical function from context to context, depending on those states.

An agent can be regarded as a state transition machine, a sort of finite state automaton (Minsky, 1972), immersed in a distributed environment, characterised by the geometrical and logical distance between the agents. When proposing a theory of an intelligent agent (Genesereth and Nilsson, 1987) described its architecture formally as a 8-tuple (we call it a GN 8-tuple):

$$GN_A =_{def} (M_A, S_A, R_A, a_A, \upsilon_A, \psi_A, \theta_A, \Phi_A)$$

where:

- M_A is the set of the agent's internals states.

- S_A is the set of all possible external states for the agent A.

- R_A is a set of partitions of S_A that characterise the agent's sensory capabilities. The

     agent can distinguish the states belonging to different partitions, but cannot distinguish the states in the same partition.

- a_A is the set of the actions the agent A can perform.

- υ_A is a sensory function that relates the states in S_A with the partitions in R_A, i.e.

$$\upsilon_A : S_A \to R_A.$$

- ψ_A is an effectory function that characterises the effects of the agent's actions on the

     external world, i.e. $\psi_A : a_A * S_A \to S_A.$

- θ_A is a function that maps an internal state and an observation into the next internal

     state i.e. $\theta_A : M_A * R_A \to M_A.$

- Φ_A is an action function that characterise the activities of the agent, determining the action of the agent whenever it finds itself in an internal state $M^i_A \in M_A$ and

recognises an external state belonging to a partition $R^i_A \in R_A$ i.e.

$$\Phi_A : M_A * R_A \to a_A .$$

This architecture, as illustrated in figure 1, was adopted by (Bandini et al, 1992) who proved its validity by experimenting it in a specific application (molecular biology).

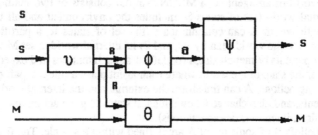

Figure 1 - The (Genesereth and Nilsson,1987) architecture for an intelligent agent

We adopted a more complex model based upon that architecture. For us, the architecture of an agent **A** is described by the following 8-tuple (we call it a MSGN 8-tuple, "Mental States Genesereth and Nilsson" 8-tuple):

$$MSGN_A =_{def} (M_A, S_A, R_A, a_A, \upsilon_A, \psi_A, BDE_A, j_A)$$

where the first six components are similar to the first six components of the architecture described above. But in our proposal the two last components are the structures BDE_A and j_A described below:

- BDE_A is a 3-tuple (b_A, d_A, e_A) where

  - b_A is a structure for the beliefs of agent A.

  - d_A is a structure for the desires of agent A.

  - e_A is a structure for the expectations of agent A.

- j_A is a structure for the intentions of agent A.

Although, there are similarities with the GN model, our own proposal supports a more advanced organisation, by adopting the idea of the Russian doll, where boxes are nested into other boxes, as it is showed in figure 2.

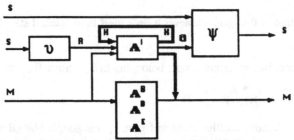

Figure 2 - The architecture organised by mental states

The world of an agent A (a MSGN 8-tuple) consists of two main parts: 1) the external world (set of states S) includes the environment and all the other agents with whom A can communicate (the set of states R, a partition of S, corresponds to the world seen by A), and 2) its internal world, a set M with four main components or mental states (beliefs, desires, intentions and expectations). Therefore, the state of the world where each agent plays a role is a pair of M and R. By doing actions, A can transform the external and the internal worlds of the other agents, and also change its own mental states (e.g. an action may generate more desires, intentions or expectations).

The beliefs B of some agent A are defined within a 4-tuple. The first part of this structure is responsible for taking into account the set of possible beliefs, including beliefs about having sets of actions, desires, intentions and expectations, and beliefs about the beliefs of other agents. The second part covers the set of possible states of belief (true and false). The third part is a function of the sets of possible beliefs into the set of states of belief, able to fix the value of a state of belief. And, the fourth part is a kind of local agent A^B (also a GN 8-tuple) responsible for the rationality of agent A, in which are defined its skills to infer, deduce and actualise its beliefs, and, in case of a conflict, to revise its beliefs, taking into account the desires and intentions of A. This sort of local agent, nested in the global agent structure, allows it to verify the abilities of agent A, i.e. it knows which actions it can or cannot do, and which actions can be executed at a certain instant in time. As a matter of fact, this local agent is a sort of active processor, within the overall architecture of the agent, in charge of implementing, in a parallel fashion, its skills and also of carrying out actions. This capability is necessary for letting the result of a reasoning chain influence the sequence of the next actions.

The desires D of an agent A are connected with the transformation of some world state i into other world state j (the desire's objective), i.e. each transformation is associated with the ideas of change and movement, and they are represented by a structure with four parts. The first one has the set of all possible pairs of world states (i and j) at a certain point in time. The second one is the set of states of desire defined by a real number associated to the notion of desire urgency and by a set of attributes: satisfied and non-satisfied desires. The third one is a function of the set of possible desires in the set of states of desire, able to fix the value of a state of desire. And, the fourth one is again a local agent A^D (a GN 8-tuple), capable of generating, at some point in time, new desires and intentions from the beliefs and desires held, or simply from the old desires. This active component is responsible for constraining the agent behaviour, depending on its preferences and desires (Sloman, 1991).

The expectations of an agent A are represented by a structure with four parts. The first part covers the domain of the expectation, the set of all possible beliefs at a certain point in time. The second one is the set of states of expectation defined by a real number associated to the notion of the expectation urgency and by the set of attributes: satisfied and non satisfied expectations. The third one is a function of the possible expectations in the set of states of expectations, at some point in time, able to fix the value of a state of expectation. And, the fourth one is again a local agent A^E (a GN 8-tuple), responsible for managing the expectations of the global agent A, and at some point in time, able to guide the agent's actions according to its beliefs. For example, an expectation that is not satisfied may generate a desire, and consequently an intention to satisfy it. The generation of the agent's plans is carried out according to its expectations of the possible world states following its actions, and, therefore, such ability allows the agent to save plans during dialogues with other agents. Expectations are generated during the interactions of the agent, just before it starts doing actions, and they correspond to the effects expected from those actions.

The intentions of agent A are also connected with a transformation from some world state i to state j (intention's objective), and are represented by a structure with four parts. The first one has the set of all possible pairs of states at a certain point in time. The second one is the set of states of intention defined by a real number associated to the notion of intention urgency and by the set of attributes: satisfied, non satisfied, waiting, active, suspended and finished intentions. The third one is a function of the possible intentions into the set of states of intentions, able to fix the value of a state of an intention. And, the fourth one is again a local agent A^I (a GN 8-tuple) for selecting an intention or a set of intentions, choosing plans and strategies to satisfy these intentions, building up contexts associated to the satisfaction of the intentions, interpreting signals from the external world according to a certain context, and suspending the course of an interaction.

3 The idea of context

An agent A interacting with the other partners, in the world, is able to perform actions, mainly motivated by some intention or by a set of intentions already existing in its mind, and these actions are linked in some sort of pattern, the so-called context. They can be communicative acts over the other partners or to be physical actions, in order to transform the state (internal or external) of the world. As a consequence, a context space is created, by relating mental states with actions, which constrain and motivate the selection of future actions.

A three-part structure accounts for the meaning of a context of interaction of an agent A, at some moment in time t, and for satisfying intention m. The first part is an ordered pair of states of the world associated to the intention m. The second part is a general strategy (Werner, 1990; 1991) for satisfying that intention. The third one is an incomplete history of the interaction till instant t and compatible with that strategy. In figure 2 the context space is represented by X, and it defines the internal states of the local agent A^I .

4 The structure of dialogue

Discussing the paper by H. C. Bunt, (Edwards and Mason, 1989) asserted that "global dialogue connectivity and control are provided by goals and beliefs represented in models maintained by the participants, and by the sequencing of dialogue acts in priority queues". Also, they put in question the ideas of dialogue control acts as context-changing functions and communicative actions as a function from contexts to contexts. Essentially, there are two different views of dialogue: the third-party observer versus the participant. This is a matter of great significance when the aim is the design of autonomous and intelligent agents immersed in some society of agents. So, the precise notion of context is required and it is critical. On account of that we defend the expedient of the agent's mental states. And, this notion of context is very important to envisage the computational tractability of the dialogues, because it contains the partial knowledge of the world and of the interactions.

An agent interacts with the world and other partners according to some particular principles, which are able to relate its mental states, the transformation of those mental states, and the actions that are effectively done by the agent. For example:

- If agent A has a belief that agent B has a desire d, and agent A has a desire to adopt the desires of agent B, then agent A adopts desire d.

- A desire may produce an intention, at some instant t, if the states of this intention are consistent with the set of agent's beliefs.

- An agent can have an intention when it believes that there is an action or a sequence of actions with which it may realise this intention.

- The interaction of the agent with the world and the other agents is caused by the concurrent behaviour of the four local agents belonging to agent's global architecture. This interactions occurs following a pattern defined by a sequence of the basic interactions (for understanding or acting) described below:

1- Agent A perceives a communicative action from agent B, and it identifies a context
for it (understanding).
2- Agent A communicates with agent B, producing a communicative action within that context or some other context (acting).
3- Agent B perceives a communicative action from agent A, and it identifies a context for it (understanding).
4- Agent B communicates with agent A, producing a communicative action within that context or some other context (acting).

These interactions are the result of the mental states transformations caused by the activities of the local agents, as illustrated in figure 3, where we can see examples of the internal process of agent A corresponding to interactions 1 and 2 above: an utterance (a communicative action) is sent from agent B to agent

A. At this point, the local agent A^I (intention) uses the component υ (sensor) for the perception and decodification of this utterance in an internal form. Along step (1), the local agent A^I associates a context (CONTEXT$_1$) to the received utterance and it produces one belief or a set of beliefs (BELIEF$_1$) that are communicated to the local agents A^B (belief) and A^D (desire). This belief (BELIEF$_1$), in agent A^D, is associated to a desire (DESIRE$_1$), and another desire (DESIRE$_2$) is produced. This particular desire, combined with a belief or a set of beliefs (BELIEF$_2$), produces an intention (INTENTION$_1$) which is communicated to agent A^I to be satisfied. At this moment, agent A^I can be processing another intention and it can attend the communicated intention only after the conclusion of this process. But, it can also interrupt and suspend this process to attend the recently received intention, depending on its urgency.

In order to satisfy intention (INTENTION$_1$), agent A^I updates the context (CONTEXT$_1$) by interacting with agent A^B. By doing so, it constructs strategies and updates beliefs (steps (4), (5) and (6)). So, agent A^I chooses a communicative action and, in the case that some expectation is associated with this action, this expectation (EXPECTATION$_2$) is included in context (CONTEXT$_1$) and is communicated to agent A^E. Afterwards, agent A^I uses component ψ (effector) of the global agent to perform the chosen action.

The role of agent A^E consists of managing expectations, i.e. it includes them, removes them and determines urgency for them. It also possible that, from an expectation which has a critical urgency, agent A^E produces a desire (DESIRE$_3$), as in step (7). In this case, a new chain of mental state transformations can be started, and this can also interrupt the running of the process which is satisfying some intention of the agent.

This chain of mental state transformations (described above and illustrated in figure 3) is the foundation for the conversations between two agents (global agents), and this overall process is related to the agent architecture by the activities of the local agents. This is quite different from (Grosz and Sidner,1986)'s approach, for whom the components of the discourse structure are properties of the discourse itself, and it is also different from the planning based approach to dialogues as, for example, (Hobbs, 1979), (Litman, 1985) and (Julien and Marty,1989).

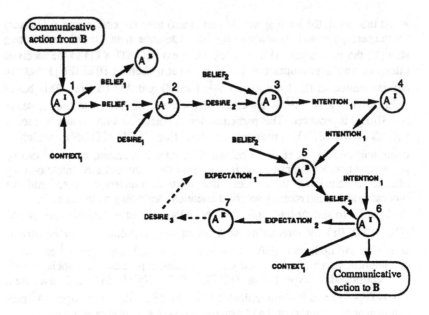

Figure 3 - Chaining of the mental states transformations by the local agents to perform conversations.

5 Evaluation of the research

We are using a criteria, based upon the guidelines described by (Cohen and Howe, 1988) for the evaluation of Artificial Intelligence research in order to validate the ideas of this research.

Our experimental work consisted of exploratory programming, and we studied and described the relationships between the different configurations of the agent's architecture, its behaviour and its environment, as it was proposed by (Cohen et al,1989) with their "behavioural ecology triangle".

Our first experiment was the development of a dialogue system based on the ideas advanced by (Power, 1974;1979). Within this system, the dialogue is produced by conversational procedures, which are lists of instructions on how utterances in pairs should be produced and interpreted. But, we verified, as (Power, 1979) had also argued, that this program was incapable of conversing flexibly, as the example of the dialogue in figure 4 shows, i.e. it was incapable of seeing a relationship between a reply of the other agent and the whole dialogue.

Figure 4 - A variation of the environment of (Power,1974;1979).

Some modifications were made introducing the memory of all the structures produced throughout the dialogue (goals, plans, candidate plans, inference rules), i.e. an embryonic notion of context was requested.

We made this kind of experiment within the same environment used by (Power, 1974;1979): two rooms and two agents, each one in one room (see figure 4). Agent A, in room 1, wants to go to room 2 and it can achieve this by moving through an open door. But, initially the door is closed and locked inside room 2 where agent B is. The agents can communicate with each other by engaging in dialogues and performing actions to satisfy their intentions.

Recently, (Draper and Button, 1990) working on development and generalising Power's work showed how his conversational procedures can be decomposed into a single type of conversational unit, the generic adjacency pair. They claimed that all speech acts and topic structures can be constructed from this unit.

This environment was also used by (Cohen,1978) to test his ideas. He was interested in showing, how deciding what to say, is a function of the belief state of the agent. Along his work, only a part of dialogue was addressed: deciding what to say in a restrict situation. He did the implementation of his ideas adopting a logic of beliefs, as a representation of the beliefs and goals of the speaker agent and of the beliefs and goals of the others.

Our second experiment was the implementation of two agents with GN 8-tuple architecture, engaged in dialogue within the same environment described above (Power's environment).

We used a language based on a logic of beliefs, in a similar way to (Cohen, 1978) to describe the internal states of the agent. In this experiment, recognition of beliefs or goals of the other participant is made in a simple way, and the agents communicate with each other adopting a restricted formal language based on logic.

Our third experiment was made by adding the notion of context (a structure with intention, expectations, beliefs, strategies and the history of dialogue), as an internal state of the agent's architecture that was employed during the previous experiment. We used also the same environment described above.

Two results were obtained. First, we proved the control of the dialogue can be obtained by managing that context. And, second, we found that the phenomena of adjacent pairs can be viewed as the result of performing communicative actions and of managing the correspondent expectations in the context.

The recognition of intentions or beliefs from an utterance of the speaker agent, by the hearer agent, is obtained by the association of a context to this utterance, or by creating a new context if this is not possible. Yet, this is still an open issue, and we are using an algorithm based on the matching of the components of the utterance with the components of the context (intention, beliefs, strategies, expectations and the history of the dialogue). When comparing our algorithm with the one proposed by (Allen, 1983) to identify a plan for an utterance there is a difference: we use more information to identify the context of an utterance.

In all these three experiments, the situation was constrained i.e. the interaction of the agent was performed inside only one context (only one plan in the cases of experiments 1 and 2). During the interaction of the agents, the change from on context to another different context was not yet possible. In order to evaluate this new situation, we made slight changes in the environment used so far. For example, during a dialogue between agents, the telephone rings, and agent A is obliged to attend it, breaking the ongoing dialogue with agent B and starting another one with the caller. And, as this internal process is caused by an external stimulus, other internal processes can be caused by internal motivation, as it is illustrated in figure 3.

In our fourth experiment we adopted another environment, a two dimensional grid (6x6) where two robots move around obstacles and try to satisfy their goals. The aim was to demonstrate that the same agent architecture, tested in the third experiment (with the context placed in the internal state) enables also the agent to navigate along a physical space to satisfy certain goals.

Although these experiments (3 and 4) are currently in progress, we have demonstrated that the GN 8-tuple is a valid way to construct an architecture for agents that either can move in a physical space or engage in dialogues.

We have been doing all this line of research upon a testbed, that was designed not only to allow controlled and repeatable experiments, but also to facilitate the creation and animation of diverse environments and agents' configurations. We have implemented it in LPA MacProlog, and the world dynamics was implemented by a simulator of coarse-grained parallelism which creates the illusion of concurrent world activity through appropriate scheduling of the agents' internal activities and actions.

6 Conclusions

Robot navigation and agent conversation are two appealing tasks in Artificial Intelligence. Both impose the co-ordination of different scientific disciplines and the involvement of several techniques. Also, the two tasks impose the need to discuss the idea of mind, and this is important for Artificial Intelligence research. Although, there are insights from one task to the other, we believe this is also relevant today for going on understanding what intelligence is.

We started our research project on modelling conversations by adopting Genesereth and Nilsson's agent model, and we proved already this model can be envisaged for navigation and conversation purposes. We were forced to introduce changes in the overall structure in order to permit more flexible and complex interactions. The main ideas were the definition of an agent architecture based upon four basic mental states and the inclusion of the notion of context. Around this kernel we put local agents acting concurrently, and capable to change these mental states and to fix the behaviour (interactions) of the agents. Therefore, the conversational contexts and the structure of agent intelligence in dialogues are defined also in terms of a subset of these mental states. So, we derived the result that the generation and the control of conversations among agents are both a consequence of the chaining of mental states transformations, in a context, according to agents' architectures. Finally, we defended a rather different definition of intention, yet closed to Werner's, by introducing links to the notions of context and architecture.

In brief, our architecture MSGN is closely associated to the theory of mental states assigned to machine intelligence. As a side effect, a theory of mental transformations supports and motivates the agent's actions. Also, the idea of associating a specific structure with an active local agent (GN) to each of the mental states allowed the study, in isolation, of the features and characteristics of each mental state, the interactions among mental states, and the concurrency among the attitudes of the mental states.

Acknowledgements

This work was partially supported by the JNICT Project STRDA/C/TIT/86/92 JURAD (supported by STRIDE and FEDER programs).

References

Allen, J. F. - Recognizing intentions from natural languages utterances, in M. Brady and R. C. Berwick (eds.), Computational Models of Discourse, MIT Press, 1983.

Allen, J. F. - Explicit models of the communication process, in Communication and Their Applications: Problems and Prospects, Trentino, Italy, November, 1990.

Bandini, S.; Cattaneo, G,; Tarantello, G. - A qualitative model of molecules as intelligent agents, University of Milan Research Report, 1992.

Bateson, G. - Mind and Nature, E.,P., Dutton, New York, 1979.

Bunt, H. C. - On formation dialogues as communicative actions in relation to partner modelling and information processing, in The Strcture of Multimodal Dialogue, M. M. Taylor, F. Neel and D. G. Bouwhuis (eds.), Elsevier, 1989.

Coelho, H. - A program conversing in portuguese providing a library service, Ph. D. Thesis, University of Edinburgh, 1979.

Coelho, H. - Facing hard problems in multi-agent interactions, Proceedings of the NATO ARW on The Future of Intelligent Systems, Sintra, October 6-10, Springer-Verlag, 1992.

Coelho, H.; Gaspar, G.; Ramos, I. - Experiments in achieving communication in communities of autonomous agents, Proceedings of the IFIP 8.3 Working Conference on Decision Support Systems, Fontainebleau, 1992.

Cohen, Paul; Howe A. - How evaluation guides AI research, AI Magazine, 9(4), 1988.

Cohen, Paul et al. - Trial by fire. Understanding the design requirements for agents in complex environments, AI Magazine, 10(3), 1989.

Cohen, P.- On knowing what to say: planning speech acts, Phd Thesis, Depatment of Computer Science, University of Toronto, 1978.

Cohen, P.; Levesque, H. - Rational interaction as the basis for communication, in Cohen, Morgan and Pollack (eds.) Intentions in Communication, MIT Press, 1990.

Draper, S; Button, C. - Conversation as planned action, planning utterances within dialogues, in Communication and Their Applications: Problems and Prospects, Trentino, Italy, November, 1990.

Edwards, J. L.; Mason, J. A. - The structure of intelligence in dialogue, in The Structure of Multimodal Dialogue, M. M. Taylor, F. Neel and D. G. Bouwhuis (eds.), Elsevier, 1989.

Galliers, J. R. - A theoretical framework for computer modules of cooperative dialogue, acknowledging multi-agent conflict, Ph. D. Thesis, Open University, 1988.

Gaspar, G. - Communication and belief changes in a society of agents, Proceedings of the Fifth Rocky Mountain Conference on Artificial Intelligence, Pragmatics in AI, Las Cruces, June, 1990.

Genesereth, M.; Nilsson, N. - Logical foundations of Artificial Intelligence, Morgan Kaufmann, 1987.

Grosz, B.; Sidner, C. - Attention, intentions and the structure of discourse, Computational Linguistics, 12(3),1986.

Hobbs, J. - Conversation as planned behavior, SRI - International, Technical Note, CA, USA, 1979.

Julien, C.; Marty, J. - Plan revision in person-machine dialogue Proceedings of the 4th ACL European Chapter Conference, Manchester, 1989.

Litman, D. J. - Plan recognition and discourse analysis: an integrated approach for understanding dialogues, Ph. D. Thesis, Computer Science Department, University of Rochester, 1985.

Litman, D. J.; Allen, J. F. - A plan recognition model for subdialogues in conversation, Cognitive Science, 1986.

Minsky, M. - Computation: finite and infinite machines, Prentice Hall, 1972.

Power, R. - A Computer model of conversation, Ph.D. Thesis, Department of Machine Intelligence, University of Edinburgh, 1974.

Power, R. - The organization of purposeful dialogues, Linguistics 17, 1979.

Rao, A.; Georgeff, M. - The formation, maintenance, and reconsideration of intentions, Proceedings of the IJCAI-91 Workshop on Theoretical and Practical Design of Rational Agents, 1991a.

Rao, A.; Georgeff, M. - Modeling rational agents within a BDI-architecture, in Allen, Fikes and Sandwall (Eds), Proceedings of the Second International Conference on Principles of Knowledge Representation and Reasoning, Morgan Kaufmann, 1991b.

Sernadas, C.; Coelho, H.; Gaspar, G. - Communicating knowledge systems, big talk among small systems, Applied Artificial Intelligence, Vol. 1, Nº 3-4, 1987.

Sloman, A. - Prolegomena to a theory of communication and affect, School of Computer Science, Birmingham University, 1991

Werner, E. - What can agents do together? A semantics for reasoning about cooperative ability, Proceedings of ECAI-90, 1990.

Werner, E. - A unified view of information, intention and ability
in Decentralized A. I., Yves Demazeau e Jean-Pierre (Eds), Elsevier Science Publishers, 1991.

Norms as mental objects. From normative beliefs to normative goals*

Rosaria Conte   Cristiano Castelfranchi

PSCS-Social Behaviour Simulation Project
Institute of Psychology, CNR
V.le Marx 15, 00137 Roma, Italy
Tel.: +39+6+86090210
Fax: +39+6+824737
rosaria@pscs2.irmkant.cnr.rm.it

Abstract. The normative dimension is essential in social life. In this work, it is claimed that a crucial aspect of the normative mechanism is the normative request, which plays a fundamental role in the spreading of normative behaviours. A social norm is here defined to be such only if it is associated with a normative request (want) that it be observed. However, within both the rational decision theory and, more recently, the multi agent systems research, norms have so far been viewed essentially as conventions. Little reference, if any, has been made to the *prescriptive* character of norms, that is, to the normative requests.

Here it is intended to draw the attention on the *prescriptive* character of norms and its role in controlling and regulating the behaviours of agents subject to them. Furthermore, a *mentalistic* notion of norms, allowing *autonomous* agents to become *normative agents* as well, is proposed. Some crucial problems concerning the mental nature of norms will be raised -- for example, how are norms represented in the agents' minds? Should they be seen as a specific mental object, and if so, which one? Which relation do they bear to beliefs and goals? Why does an autonomous agent comply with norms? Some initial solutions will be proposed.

1   Introduction

The study of norms is a recent but growing concern in the AI field (cf. [MOS92]; [SHO92]) as well as within other formal approaches to social action, such as Game Theory (where theories of norms began to appear long before, cf.[LEW69]; [ULL77]). The normative is an essential dimension of social life. An explicit model of such dimension is crucial for any theory of social action. Notions such as "social commitment" (cf. [GAS91]; [COH90]; etc.) "joint intentions", "teamwork" [COH91], "negotiation" [DAV83], "social roles" [WER89] etc. will not escape a purely metaphorical use if their normative character is not accounted for. Indeed, *one's committing oneself* before someone else causes the latter to be *entitled*, on the very

---

* A preliminary version of this paper has been presented at the '93 AAAI Spring Symposium Series on "Reasoning about mental states: Formal theories and applications"

basis of commitment, to expect, control, and exact the action one has committed oneself to.

Furthermore, there is no true *negotiation* without *control for cheaters* and a *norm of reciprocation* (cf. [GOU60]).

Analogously, joint intentions and teamwork arise from true negotiation and create *rights* based on commitments. Finally, roles are but sets of *obligations* and *rights*.

A *norm-abiding behaviour* need not be based on the *cognitive processing of norms* (it might be simply due to imitation). On the other hand, a normative decision-making does not necessarily produce a behaviour conforming to norms (some norm operates in the decider's mind even if the final decision leads to a transgression). To model the normative reasoning does not imply to model a norm-abiding system. However, a large portion of the autonomous agents' normative behaviour is allowed by a cognitive processing of norms. Therefore, norms are not only an essentially *social* but also a *mentalistic* notion. Norms are indeed a typical *node* of the *micro-macro link*, of the link between individual agents and collective phenomena, minds and social structures. In line with [SHO92], we maintain that an epiphenomenal view of normative systems, as "emerging properties" of the social systems improving co-ordination within the social systems themselves, (cf. [MOS92]) is not sufficient.

The emergence paradigm gives an account of conventional norms and conforming behaviours (cf. [BIC90]), but leaves unexplained the *prescriptive character of norms*. Within the game-theoretic paradigm, a social norm is defined as an *equilibrium*, a combination of strategies such that "each maximises his expected utility by conforming, *on the condition that* nearly everybody else conforms to the norm." (cf. [BIC90: 842]) Social norms are defined as behavioural regularities which emerge from the strategic agents' choices to conform. This is an account of the spreading of certain behaviours over a population of strategic agents, in a word, a model of social conventions. However, such a model does not account for a crucial aspect of the normative mechanism, which plays a role in the spreading of normative behaviours, namely the normative request. What is lacking in the game-theoretic definition aforementioned is a reference to each *agent wanting others to conform to the norms*. A social norm, indeed, is such if it is associated with the normative want that it be observed.

Therefore, here it is intended to draw the attention on both the following aspects:

• the *prescriptive* character of norms, that is, their role in controlling and regulating the behaviours of agents subject to them;

• a need for a *mentalistic* notion of norms allowing cognitive agents to become normative agents as well. Norms are not yet sufficiently characterised as mental objects (CF [SHO92]). Here, some crucial problems concerning the mental nature of norms will be raised and some initial solutions proposed.

2 The normative coin.

What is a norm in an agent's mind? This question leaves aside another fundamental question, namely what is a norm tout court. Although prioritary, we will not face this question here (however, see [CAS91]).

Here, we will focus exclusively on the "internal" side of norms, that is, What cognitive role do norms play in the agents' minds and in which format are they represented? What kind of mental object is a norm? What is a normative authorship in the agents' beliefs?

There are at least two distinct ways in which *norms can be implemented* on a computer system: as *built-in* constraints (like production rules) or as explicit and *specific mental objects* (i.e. obligations, duties, etc.) distinct from, say, goals and beliefs. Of course, there might be other solutions: for example, and in line with the game-theoretic view, one might think of noms as a way of describing the *behaviours of strategic agents in interaction* with no need for hypothesising a specific, either mental or social, normative object. As already said, this is an epiphenomenal view of norms, which essentially nullifies them. An intermediate alternative consists of implementing norms just as goals, namely *final ends*: In this case, cognitive agents would be allowed to choose among their (competing) goals (instead of simply applying procedures and routines) but they would treat norms as any other goal of theirs. At least two undesirable consequences seem to derive from this alternative: one agent observing the norms would come to depend exclusively upon that agent's subjective preferences, and there would be no social control and influencing. To implement norms as specific mental objects, although a costly alternative, is required by a *cognitive modelling* approach to social agenthood, in which the present work is framed. Moreover, the apparatus of regulation of the human species has in fact developed a cognitive representation and processing of norms. Undoubtedly, the species has profited by such an explicit representation of norms. A crucial but difficult task, beyond the scope of the present work, would be to explore these advantages and confront them with those of all alternatives considered (however, for initial suggestions, see [CON94a]). Here, instead, we will take for granted the advantages of a cognitive representation of norms, and limit ourselves to explore the ingredients of such a representation.

Our language draws on [COH90]'s model for describing rational interaction. For the readability of the paper, let us provide the semantics of the main predicates used and a rule for means end reasoning (GGR) here called goal-generation rule (for further analysis, see [COH90], [CON91a], [CON91b] and [CON95]):

a:	action;
p and q:	states of the world;
v:	a world state positively valued;
x, y, and z	single agents;
$(BEL\ x\ p)$:	x has p as a belief
$(GOAL\ x\ p)$:	x wants that p is true at some point in the future.
$(OUGHT\ p)$:	an obligation concerning any given proposition p;
$(DONE\ x\ a) = def\ (DONE\ a) \wedge (AGT\ x\ a)$	
	action a has been done by agent x;
$\Diamond p$	p is true at some point in the future.

$$((GOAL\ x\ p) \wedge (BEL\ x\ (q \supset p))) \supset (GOAL\ x\ q)$$

$$(GGR)$$

189

if x wants p and believes that if q than p will follow, then x will want that q as well.

$(OBTAIN\ x\ p) =_{def} (GOAL\ x\ p)\ \wedge \Diamond\ (p\ \wedge\ (BEL\ x\ p))$

x obtains p iff p is a goal of x's and later it will be true and x will believe so.

$(GOAL\text{-}CONFL\ x\ y\ p\ q) =_{def} (GOAL\ x\ p)$

$\wedge\ (GOAL\ y\ q)$

$\wedge\ (p \vee q)$

goals which consist of incompatible propositions are in conflict.

3 The normative belief.

Norms are generally considered as prescriptions, directives, commands. Even when expressed in any other type of speech act they are meant to "direct" the future behaviour of the agents subject to them, their addressees (As).

To this end, they ought to give rise to some new goal in an addressee's mind (cf.. [CON91]): for autonomous agents to undertake (or abstain from) a course of action, it is not sufficient that they know that this course of action is wanted (by someone else) (cf. [ROS88, GAL90]). It is necessary that these agents have the goal to do so. Norms may act as a mechanism of goal-generation. Indeed, they represent a powerful mechanism for inducing new goals in people's minds in a cognitive way. How is this possible?

At first, norms shall be represented as beliefs in an A's mind. Let us start from beliefs about requests. A simple request, before succeeding and producing acceptance on the side of the recipient, is nothing but a belief, namely a belief about someone's want. Such a belief can be expressed as follows:

$(BEL\ x\ (GOAL\ y\ (GOAL\ x\ p)))$ (1)

where agent x believes that someone else, agent y, wants x to have the goal that p. More specifically, x believes that what y requires of her is to do an action planned for p:

$(BEL\ x\ (GOAL\ y\ ((DONE\ x\ a)))$ (2)

Now, two questions arise here:

1. what is the difference, if any, between this type of belief, that is, a belief about an *ordinary want*, and a belief about a normative or *prescriptive want*? Is a norm always represented in people's minds as an expression of some particular external want?

2. how do we go from a belief about a normative want to the goal to comply with it, that is, to a normative goal?

Starting from the former question, one could say that normative beliefs are beliefs about a general want affecting one or more agents. This equals to saying that there is a class of agents wanting some of them to accomplish a given action. This view is interesting and fits rather well the DAI field, where attention is now increasingly paid to shared mental states and collective action ([GRO90, LEV90, COH91], etc.; for review and further analysis, see [RAO92]). However, it presents two drawbacks.

First, it relies upon a notion of collective or group's want as yet fundamentally distributive. In [RAO92], for instance, a social agent's wants and beliefs are defined as the *conjunction* of goals and beliefs of the group's members. The authors contrast this with the opposite view (held, for instance, by [SEA90]) of social entity as "irreducible" to their members. An alternative to both views exists, which consists of saying that agents form a collective if *they objectively depend on one another to achieve one and the same goal or interest* (the latter being defined in [CON94b] as a world state not necessarily wanted nor believed by the agents involved that nonetheless implies the future achievement of (one of) their goals). Furthermore, the collective may achieve its common goal, or realise its common interest, by distributing tasks among its members. However, the task assignment may be accomplished by a specialised sub component of the group. In other words, the notion of *group* or collective *want* can be "reduced" to the mental states (of some) of its members without necessarily implying shared goals nor mutual beliefs, In this sense, it represents an intermediate solution between the first two views.

Secondly, and moreover, a normative belief seems to be grounded on something more than a general request. In particular, what seems to be implied is a notion of *obligation*, or duty, as distinct from the classical deontic notion of necessity. The present notion of obligation is very similar to that proposed by Jones and Porn (see [JON91]; for a discussion, see [CON95]), which does not imply that an obligatory proposition p be true in all the words where the obligation is in force. Therefore, a normative belief refers to this notion of obligation, rather than to the classical notion of necessity. A normative belief, in other terms, may refers to some proposition that, in some sub-ideal worlds, is not verified.

Let us express the general form of a normative belief as follows:

$$(N\text{-}BEL\ x\ y_i\ a) =_{def} (\Lambda_{i=1,n}(BEL\ x\ (OUGHT\ (DONE\ y_i\ a))))) \qquad (3)$$

where $(OUGHT\ (DONE\ y_i\ a))$ stands for an *obligation, in the sense previously defined, for a set of agents y_i to do action a*. The question is: what relation does (3) bear with belief type (2)? This relation seems possible thanks to the notion of *normative request*. In other terms, a normative belief implies a belief about the existence of a normative request:

$$(N\text{-}BEL\ x\ y_i\ a) \supset \exists z (BEL\ x\ (GOAL\ z\ (OUGHT\ (DONE\ y_i\ a)))) \qquad (4)$$

where z belongs to a higher level set of agents y_j of which y_i is a subset $(V_{j=1,q>n}$ $(z=y_j)$): An agent x has a normative belief about action a if x believes that *someone wants* it to be *obligatory* for y_i to do a. This is a minimal condition, since z might be a set of agents. It might even coincide with the whole superset y_j. In the latter case, the normative request overlaps with the group's want. But this is not necessary. Suffice it to say that, in a normative belief, a sub component (individual or social[1])

---

[1] As in [RAO92], the definition of a social entity should be recursive. Therefore, what has been said with regard to the group at large applies to its

of the group is mentioned to issue a request. In x's belief, a request is normative on condition that a given z is believed to want y_i to have an obligation to do a.

To be noted, unlike that of deontic necessity, the notion of obligation we are referring to allows not only for transgressions and exceptions, but also for normative requests to be distinguished from coercive requests. In coercion, the coercive agent does not want the victim to believe in an abstract obligation to do something. All he needs is to persuade the victim that she is *forced* (in the sense of necessity) to do *what the coercive agent wants*. The victim must believe that she is left with no alternative, no choice. On the contrary, a normative request is believed to create a mental state of abstract obligation, independent of any real necessity. By default a normative request is one which wants you to have an obligation to do something, and not simply the belief that you could not do anything else.

Furthermore, a normative request is usually believed to be *grounded on norms*, to be norm-based. More specifically, a strong sense of normative request occurs when the source of the request is characterised as a normative authorship, *held to* issue norms:

$$(N\text{-}BEL\ x\ y_i\ a) \supset \exists z (BEL\ x (OUGHT(GOAL\ z\ (OUGHT(DONE\ y_i\ a))))) \qquad (5)$$

A weaker or milder meaning is that of entitled, or *legitimate* will, a legitimate goal being defined as follows:

$$(L\text{-}GOAL\ x\ p) =_{def} \forall y \exists q (GOAL\text{-}CONFL\ x\ y\ p\ q) \supset (OUGHT\ \neg(GOAL\ y\ q)) \quad (6)$$

in words: p is a legitimate goal of x's iff forall agents y that happen to have a given goal q conflicting with p, y ends up with having an obligation to give up q. In other words, a legitimate goal is watched over by a norm. In this sense, rights and legitimacy are said to give assistance to, and go to the rescue of, those who cannot defend themselves against aggression and cheat. Consequently, an entitled goal is that which a norm protects from any conflicting interest.

To sum up, an agent has a normative belief iff that agent believes that there is an obligation for a given set of agents to do a given action. At a more careful examination, however, the obligation is believed to imply that:

- a given action is *prescribed*, that is, requested *by*

- *a norm-based will*, be it held to issue that request, or simply entitled to do so.

Of course, a normative belief does not imply that a deliberate issuing of a norm has *in fact* occurred. Social norms are often set up by virtue of functional unwanted effects. However, once a given effect is believed to be a social norm, an entitled will is also *believed* to be implied, if only an anonymous one ("You are wanted/expected to (not) do this...", "It is generally expected that...", "This is done so...", etc.).

This equals to saying that the present model of normative beliefs is recursive. A request is believed to be normative if it ultimately traces back to some norm. This is not to say that norms are "irreducible" objects. Of course, the origins of norms call for an explanation which unavoidably brings into play the community of agents, their interests and their interactional practice (cf. [ULL77]). However, in the agents'

---

subcomponents as well. In case z in turn is a multi agent subcomponent, its will might be shared or not among its members. Its task (say, to legiferate) might be accomplished in such a way that not all members share the same goals.

representations there is no need for keeping a record of such history. In the agents' beliefs a norm is always represented as a legitimate, even a norm-driven, prescription. The present model tries to give an account of this evidence.

4 The route of norms in the mind.

Turning to question 2. raised above, a normative belief is only one ingredient of the normative reasoning. *Norms, indeed, are hybrid configurations of beliefs and goals.* Actually, as defined so far, a normative belief is only descriptive: it does not "constrain" the believer and his decisions. Indeed, an observer's description of a society's rules does not influence in any relevant way her decisions. What is needed for an agent to regard herself as subject to, addressed by, a given norm?

The pertinence belief. First another belief is needed, namely a pertinence belief: For x to believe that she is addressed by a given norm, x needs to believe that she is a member of the class of As of that norm:

$$(P\text{-}N\text{-}BEL \ x \ a) =def \ (\Lambda_{i=1,n} \ (N\text{-}BEL \ x \ y_i \ a)) \wedge (V_{j=1,n} \ (BEL \ x \ (x = y_j))) \qquad (7)$$

where *P-N-BEL* stands for normative belief of pertinence.

Now, x's beliefs tell her not only that there is an obligation to do action a, but also that the obligation concerns precisely herself.

The normative goal. Still, *(7)* is not much less "descriptive" than *(3)*. We do not see any normative goal, yet.

First, let us express a normative goal as follows:

$$(N\text{-}GOAL \ x \ a) =def \ (P\text{-}N\text{-}BEL \ x \ a) \wedge (GOAL \ x \ (DONE \ x \ a)) \qquad (8)$$

A normative goal of a given agent x about action a is therefore a goal that x happens to have as long as she has a pertinence normative belief about a. Ultimately, x has a normative goal in so far as and because she believes to be subject to a norm. Therefore, a normative goal differs, on one hand, from a simple constraint that reduces the set of actions available to the system (cf. [SHO92]), and, on the other, from other ordinary goals.

With regard to behavioural constraints, a normative goal is less compelling. An agent endowed with normative goals is allowed to compare them with other goals of hers and to some extent freely choose which one will be executed. Only if endowed with normative goals an agent may legitimately be said to comply with, or violate, a norm. Only in such a case, indeed, she may be said to be truly normative.

With regard to ordinary goals, a norm-goal is obviously more compelling: when an agent decides to give it up, she knows she both thwarts one of her goals and violates a norm[2].

Now, the question is: How and why does a normative belief come to interfere with x's decisions? What is it that makes her "responsive" to the norms concerning her? What is it that makes a normative belief turn into a normative goal?

---

[2] Intuitively, she gives up both the expected consequences of the action prescribed (any worldstate supposedly convenient to the agent or otherwise positively valued) *and in addition* sustains the costs of N-transgression. Although required, a formal treatment of both aspects is beyond the scope of this work.

Goal- and norm-adoption. There seem to be several ways of accounting for the process leading to normative goals (N-goals) as well as several alternative ways of constructing a normative agent. There also seems to be a correspondence between the process from a belief about an ordinary request to the decision of accepting such a request, which we called (cf. [CON91b]) *goal-adoption*, and the process from a N-belief to a N-goal, which by analogy will be called here *norm-adoption* (see Table 1).

	Goal-Adoption	Norm-Adoption
1. C o n d i t i o n a l Action	**Slavish** (BEL x (GOAL y (DONE x a))) ⊃ ◊ (DONE x a)	**Automatic** (P-norm-BEL x a) ⊃ ◊(DONE x a)
2.Instrumental Adoption thanks to (GGR)	**Self-interested** ∀p∃q(BEL x ((OBTAIN y p) ⊃ ◊(OBTAIN x q)))⊃ ◊(GOAL x (OBTAIN y p))	**Utilitarian** ∀a∃p((P-norm-BEL x a) ∧ (BEL x ((DONE x a)⊃ ◊(OBTAIN x p))))⊃ ◊(norm-GOAL x a)
3.Cooperative Adoption thanks to (GGR))	**Co-interested** ∀p∃q(BELx((OBTAINy p)⊃ ◊q))⊃ ◊(GOAL x (OBTAIN y p)) with .q's being (in x's beliefs) commonly wanted by x and y	**Value-driven** ∀a∃q((P-norm-BELx a)∧ (BEL x ((DONE a) ⊃ ◊q)))⊃ ◊(norm-GOAL x a) with q's being (in x's beliefs) a world state positively value by both x and the normative source: (BEL x (q = v_{(x z)})) with v standing for any value.
4. T e r m i n a l Adoption	**Benevolent** $(\Lambda_y=1,n$ (GOAL x (OBTAIN y p_y))) with p_y being the set of y's goals.	**Kantian** $(\Lambda_x=1,n$(norm-GOAL x n_{(x)})) with a_x being the set of norm-actions required of x

Table 1: **The route of norms in the mind.**

In situation 1. (**conditional action**), we find some sort of production rule: in goal-adoption (G-A), anytime a request is received by a system endowed with such a rule, a goal that *a* be done is fired. Analogously, in norm-adoption (N-A), anytime a N-belief is formed a N-goal is fired. Now, this is a rather cheap solution: *no* reasoning and *autonomy* are allowed. It is simple machinery that could be of help in cutting short some practical reasoning, but is insufficient as far as the modelling of normative reasoning is concerned. However, such a rule seems to account for a number of real-life situations. Think, as far as *slavish* G-A is concerned, of the habit of giving instructions when asked by passengers, and in the case of *automatic* N-A, of the routine of stopping at the red light (of course, in situations 1, it is hard to differentiate G-A from N-A).

In situations 2 (**instrumental adoption**), *greater autonomy* is allowed: adoption is subject to restrictions. In G-A, on the base of this rule, *x* will *self-interestedly* adopt only those of y's goals that *x* believes to be a sufficient condition for *x* to achieve some of her goals. Typically, but not exclusively, this rule depicts situations of *exchange*. An *utilitarian* N-A rule says that forall norms, *x* will have the corresponding N-goals if she believes she can get something out of complying with them. (Think of the observance of norms for fear of punishment, need of approval, desire to be praised, etc..)

Cooperative, or *co-interested*, goal adoption occurs whenever an agent adopts another's goal to achieve a common goal. N-adoption is cooperative when it is *value-driven,* that is, when the agent autonomously shares both the end of the norm and the belief that the latter achieves that end. This type of N-A can be seen as some sort of moral cooperation since the effect of the norm is shared (in the N-addressee's beliefs) by the addressee and the normative source.

The last situation is **terminal adoption**. This is not a rule, but a *meta-goal* which is defined, in the case of G-A, as *benevolent* (*x* is benevolent with regard to *y* when she wants the whole set of y's goals to be achieved), and, in the case of N-A, *"Kantian"* "(*x* wants to observe the whole set of norms addressing herself as ends in themselves).

In situation 1, the rule is a typical production rule. Its output is an *action*. In situations 2 and 3, the rules give rise to some specific *goals*. In the case of N-A, the agent ends up with a new type of goal, namely a *normative goal.*

As seen at the beginning, this implies *x*'s belief that she is requested to do *a* by a normative will. But it implies two further beliefs as well, namely that the normative source is *not acting in its own personal interests;* and that *other agents are subject to the same entitled request* (in a normative belief, a set of N-addressees is always mentioned). Now, these further aspects play a relevant role, especially within the process leading from N-goals to N-actions.

A N-goal, in fact, is not sufficient for an agent *to comply with* a norm. Several factors occurring within the process leading from N-goals to N-actions might cause the agent to abandon the goal and violate the norm. Among the others (more urgent conflicting goals; low expected chances of being caught red-handed, etc.), what is likely to occur is a confrontation with other addressees of a given norm. As known, a high rate of transgressions observed discourages one's compliance. Viceversa, and for the same reason, it is possible to show that if one has complied with a given norm, one will be likely to influence other agents to do the same (normative equity). Indeed, it can be argued (cf. [CON92]) that *normative influencing* plays a rather relevant role in the spreading of normative behaviour over a population of autonomous agents.

5 Conclusive remarks and future research

In this paper, the necessity of a cognitive modelling of norms has been argued. It is proposed to keep distinct the normative choice from any norm-like behaviour, that is that behaviour which appears to correspond to norms. Such a difference is shown to be allowed only thanks to a theory of norms as a two-fold object (internal, that is, mental and external, or societal).

Some instruments, still rather tentative, have been proposed for a formal treatment of the "internal side" of norms. In particular, a view of norms as a complex mental object has been attempted. This object has been shown to consist of other more specific ingredients, namely goals and beliefs. Two notions of normative belief and goal have been provided and discussed, and aspects of the process of N-adoption examined and confronted with the process of adopting ordinary goals.

Acknowledgements

We would like to thank both Gianni Amati, for his careful reading and commenting the paper and providing solutions, and Amedeo Cesta for his helpful suggestions.

References

[BIC90]   Bicchieri, C Norms of cooperation. *Ethics, 100*, 1990, 838-861.
[CAS91]   Castelfranchi, C. & Conte, R. Problemi di rappresentazione mentale delle norme. Le strutture della mente normativa, in R. Conte (ed.) *La norma. Mente e regolazione sociale*. Roma, Editori Riuniti, 1991, 157-193.
[COH90]   Cohen, P. R. & Levesque, H. J. Persistence, Intention, and Commitment, in P.R Cohen, J. Morgan & M.A. Pollack (eds.) *Intentions in Communication*. Cambridge, MA, MIT, 1990.
[COH91]   Cohen, P. & Levesque, H.J. *Teamwork.*. TR-SRI International, 1991.
[CON91]   Conte, R., Miceli, M. & Castelfranchi, C. Limits and levels of cooperation. Disentangling various types of prosocial interaction. In Y. Demazeau & J.P. Mueller (eds.) *Decentralized AI-2*. North-Holland, Elsevier, 1991, 147-157.
[CON92]   Conte, R. & Castelfranchi, C. Minds and Norms: Types of normative reasoning. In C. Bicchieri & A. Pagnini (eds.), *Proceedings of the 2nd Meeting on "Knowledge, Belief, and Strategic Interaction"*, Cambridge University Press, in press; TR-IP-PSCS, 1992.
[CON94a] Conte, R. Norme come prescrizioni: per un modello dell'agente autonomo normativo. *Sistemi Intelligenti, 1*, 1994, 9-34.
[CON94b] Conte, R. & Castelfranchi, C. Mind is not enough. Precognitive bases of social action. In N. Gilbert & J. Doran (eds.), *Simulating Societies: the computer simulation of social processes*, London, UCL Press, 1994, 267-286.
[CON95]   Conte, R. & Castelfranchi, C. *Cognitive and social action*. London, UCL Press, 1995.
[DAV83]   Davies, R. & Smith P.G. Negotiation as metaphor for distributed problem-solving. *Artificial Intelligence, 20*, 1983, 63-109.

[GAL90]   Galliers, J.R. The positive role of conflict in cooperative multi-agent systems, in Y. Demazeau, & J.P. Mueller (eds) *Decentralized AI*. North-Holland, Elsevier, 1990.

[GAS91]   Gasser, L. Social conceptions of knowledge and action: DAI foundations and open systems semantics. *Artificial Intelligenc, 47*, 1991, 107-138.

[GRO90]   Grosz, B.J. & Sidner, C.L. Plans for discourse, in P.R. Cohen, J. Morgan, & M.E. Pollack (eds.) *Intentions in communication*. Cambridge, MA, MIT Press, 1990.

[GOU60]   Gouldner, A. The norm of reciprocity: A preliminary statement. *American Sociological Review, 25,* 1960, 161-179.

[JON91]   Jones, A.J.I. & Porn, I. On the logic of deontic conditionals, in J.J.C Meyer & R.J. Wieringa (eds.) First International Workshop on Deontic Logic in Computer Science, 1991.

[LEV90]   Levesque, H.J., Cohen, P.R., & Nunes, J.H.T. On acting together. *Proc. of the Eighth National Conference on Artificial Intelligence (AAAI-90),* 1990, 94-99.

[LEW69]   Lewis, D. *Convention*. Cambridge, MA, Harvard University Press, 1969.

[MOS92]   Moses, Y. & Tennenholtz, M. On Computational Aspects of Artificial Social Systems. *Proc. of the 11th DAI Workshop*, Glen Arbor, February 1992.

[RAO92]   Rao, A. S., Georgeff, M.P., & Sonenmerg, E.A. Social plans: A preliminary report, in E. Werner & Y, Demazeau (eds.) *Decentralized AI - 3*. North Holland, Elsevier, 1992.

[ROS88]   Rosenschein, J.S. & Genesereth, M.R. Deals among rational agents, in B.A. Huberman (ed.) *The ecology of computation*. North-Holland, Elsevier, 1988.

[SEA90]   Searle, J.R. Collective intentions and actions, in P.R. Cohen, J. Morgan, & M.E. Pollack (eds.) *Intentions in communication*. Cambridge, MA, MIT Press, 1990.

[SHO92]   Shoham, Y. & Tennenholtz, M. On the synthesis of useful social laws for artificial agent societies. *Proc. of the AAAI Conference*, 1992, 276-281.

[ULL77]   Ulman-Margalit, E. *The emergence of norms*. Oxford, Clarendon, 1977.

[WER89]   Werner, E. Cooperating agents: A unified theory of communication and social structure, in M. Huhns and L. Gasser (eds.), *Distributed artificial intelligence, Vol. 2*, Kaufmann and Pitman, London, 1989.

Multi-agent architectures

The hedonic agent : a constructivist approach of abductive capacities

Paul Bourgine

AL & AI lab., CEMAGREF, Parc de Tourvoie, 92185 Antony, France
Tel : 33 1 40966179, Fax : 33 1 40966080

Abstract. The most important question that autonomous agents have to answer is how to remain viable in various and changing environments despite their bounded cognitive capacities. This question is thus the same as how their semiotic capacity to guess viable solutions emerges, that is abduction. The claim is that no learning can happen without a hedonic principle. That defines the hedonic level.

The hedonic level is presented as a cognitive paradigm : the hedonic agent can auto teach its hedonic and sensorimotor anticipations and also the meaningful and useful distinctions for these anticipations. That defines the possibility of the emergence of a job architecture, in a constructivist way.

A model of emergence of abductive capacities inside an architecture of jobs and inside jobs is proposed. This model takes into account both the limited cognitive capacities of the agent and its necessity to manage continuously its compromise between exploration and exploitation. The claim is that, inside its job architecture, the hedonic agent can use only forward policies because of its bounded cognitive capacities. The theory of bandit processes provides the optimality of such policies based on the index of Gittins and their pertinence for the compromise between exploration and exploitation. A new learning rule of reinforcement, the I-Learning rule, is proposed to evaluate this index.

1 Introduction

The most important question that autonomous agents have to answer is how to remain viable in various and changing environments. This question is thus the same as how emerges the semiotic capacity to guess viable solutions, that is abduction.

Abduction is one of the most important property of autonomous systems. It is the capacity to interpret some signs of a situation to find a viable solution. This concept was first proposed by Peirce, a mathematician and philosopher of last century, who has proposed the bases of semiology. He postulates that abduction is a property of animals, that is their common mode of "reasoning".

The necessity to postulate abduction is closely related to idea of bounded cognitive capacities. More bounded the cognitive capacities are the more direct capacities to guess viable actions are required.

The class of autonomous agents considered here is those which have to perform some complicated jobs in various and changing environments to maintain their viability despite their bounded cognitive capacities. The aim of this paper is to sketch a theory of such autonomous agents. The attempt here is to model them in the same framework, no matter if they are natural or artificial like usually in the cybernetic tradition.

It would be difficult to think that simple reactive agents can perform many complicated tasks in various environments, especially if the environment is continuously changing. Even if every autonomous agent presents a basic repertory of simple reactive strategies for simple jobs, it has to build by learning from experience a more sophisticated repertory of more complicated jobs in various and changing environments.

A first thesis is that no learning can happen without an hedonic principle. Without such principle everything should be equally learned which is a huge and impossible task. This thesis defines the hedonic level.

A second thesis here is that all abductive capacities emerge by auto teaching in a constructivist way, i.e. by a kind of learning which takes place only inside the cognitive system of the agent.

A third thesis is that the policy of an agent with bounded cognitive capacities is necessarily a forward policy in its forward historical process.

The most important issue with auto teaching by experience is the permanent compromise between exploration and exploitation. The permanent question for the hedonic agent in various and changing environments is : when it is better to continue to exploit a well known job ? or to explore an alternative less known job ? This question is treated by taking inspiration from the theory of bandit process [e.g. Gittins, 89].

The hedonic level takes place inside a new emergent cognitive paradigm which goes from the constructivism of Piaget to the enactivism of Varela [Varela, 86], including the approach of adaptation [Holland, 75], categorisation [Rosh, 78] primary consciousness [Edelman, 92] and abduction [Bourgine & Varela, 92]. The main ideas of this new cognitive paradigm are the following : the mind is embodied ; cognition is enaction of a world of meanings through the history of the coupling between the agent and its environment. The criterion of success is the viability of the autonomous agent. The main properties of agent are situatedness and abduction. This new cognitive paradigm is very present in the new domains of Artificial Life [Langton, 89, 92] [Varela & Bourgine, 92], and the nouvelle AI [Brooks, 91].

The first part of the paper explicates the hedonic agent as a cognitive paradigm : the hedonic agent is considered with both its reactive level and its hedonic one, eventually completed by an eductive one. The second part proposes a standard model for a hedonic agent by treating the issue of the compromise between exploitation and exploration.

2 The hedonic agent as a cognitive paradigm.

The class of autonomous agents considered here is the agents which can adapt themselves to various and changing environments by auto teaching from experience. The most important case is sensorimotor systems, in a broad sense, no matter if they are natural or artificial. The language of sensorimotor autonomy is used in this part for this reason and for the sake of clarity.

One of the most remarkable fact of biological evolution is that the co-evolutionary process has produced more and more powerful cognitive capabilities. What follows is a proposition to categorise sensorimotor autonomy in three levels by taking inspiration in Nature. At first animals have essentially a reactive behaviour. That defines a first level of sensorimotor autonomy. With the development of nervous systems, the possibility to correlate internal pleasure with the sensorimotor state leads to hedonic behaviours by self-reinforcement mechanisms, which define the hedonic level. And the possibility to auto teach its own sensorimotor dynamics through a huge number of sensorimotor cycles leads to the eductive level, which consists of sensorimotor anticipations. All these levels only take into account the internal cognitive state of a sensorimotor system.

These levels are presented in an increasing cognitive capacity. Therefore each level inherits from the most important features of the preceding ones. The eductive level adds new capacities to the hedonic one which adds new capacities to the reactive one.

But the aim here is that all these levels can be together present and co-operating in the same hedonic agent. The hedonic level is the fundamental cognitive level considered in this part.

2.1. The reactive level

Reactive systems are sensorimotor systems equipped to produce directly the pertinent distinctions and to perform directly some excellent action for their viability. For example the bee has a vision of the UV light ; and when a field of flowers is observed with a UV filter, all is uniformly white except the pistils which are black. Therefore it is easy for a bee to fly towards its source of food. Indeed the neural network of insects is composed of a restraint number of neurones. And the neurones have generally a very specific role like in a clock maker's mechanism.

Thus, we have to postulate that a reactive system has a strategy function, which is implemented for example by a recurrent network with constant weights. And its dynamics provides reactively the next action from its internal perceptions, which are of three kinds : the proprioception of what it is doing ; the exteroception of what it is looking at its environment ; and the interoception of what it is perceiving inside its self organised processes (fig 1).

The strategy function contains all the abductive capacity of the system. We have to postulate that this strategy allows sufficient adapted actions to maintain internal self-organised processes inside their long term viability domain. Without that, the system could not reach a normal life length. We can summarise this discussion in the following characterisation :

(the reactive autonomy hypothesis)

a- An autonomous reactive agent is a sensorimotor system which performs its actions with a strategy function through a fixed categorisation of its perceptions.

b-The type of the sensorimotor system and the shape of the strategy function give to the agent a sufficient abductive capacity in its usual environments to keep the system viable for a normal life time.

This hypothesis is convenient only for a very basic repertory of jobs. The fixed character of the strategy doesn't allow the agent to adapt itself to moving environments or to complicated jobs.

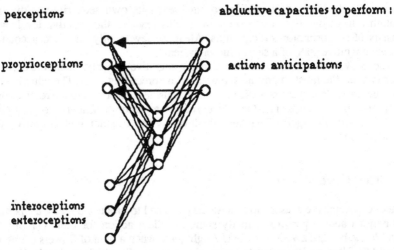

fig 1 : the reactive agent has a cognitive system which has abductive capacities to perform viable actions from its perceptions. The cognitive system is a recurrent network (the back arrows) whose innate dynamics providing a fixed strategy. It has an innate categorisation capacity which is adequate and pertinent for anticipating useful actions : that is the role symbolised by the hidden intermediate layer(s).

2.2 The hedonic level

Now the autonomous agent doesn't have a constant behaviour. That is necessary if its environment changes. The question is now how it can learn new strategy functions. The fundamental idea is that no learning is possible without a pleasure principle. Without a pleasure principle, all everything has to be learned in an equinomious way, which is impossible. Thus the simplest hypothesis is to postulate a hedonic function (even very sophisticated) and a metadynamics to auto teach hedonic anticipations. Then the agent can revise its strategy function according to the only things it knows to learn, i.e. its hedonic anticipations.

Such hypothesis can be grounded on neuroscience. In learning mechanisms studies, the focus is generally on the synaptic efficiency modification. But the central nervous system of vertebrates synthesises a huge number of chemical substances. And the global rewarding or punishing properties of the substances involved in pain, pleasure, fear, etc. provide a general learning mechanism by self reinforcement. All these feelings are related to internal homeostatic viability constraints and to external viability constraints related to dangerous situations.

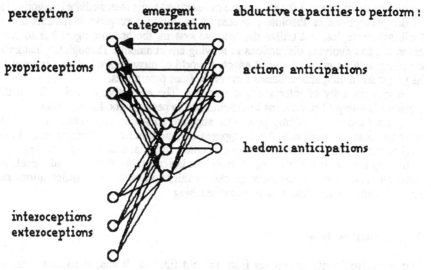

fig 2 : the hedonic agent has a cognitive system whose first metadynamics auto teach by self reinforcement from the internal reward to anticipate the hedonic value of its sensorimotor state. A second metadynamics allows to auto teach hedonic strategies. All the auto teaching processes supposes a third metadynamics which allows the agent to acquire emergent categorisation which is adequate and pertinent for its useful anticipations : that is the role symbolised by the hidden intermediate layer(s).

Furthermore, if we admit that all these feelings are more or less transgressable and interacting constraints, we can aggregate them into a global pleasure measure. The hedonic function represents a synthesis of the internal and external viability constraints.

(the hedonic autonomy hypothesis)

a- An autonomous hedonic agent is a sensorimotor system with a hedonic function. In a very slowly moving environment, it can auto teach, through an emergent categorisation of its internal perceptions, hedonic anticipations by actively experimenting its possible strategies.

b-The shape of the hedonic function represents sufficiently the internal and external viability constraints to give to the agent a sufficient abductive capacity in its usual slowly evolving environments.

When the perceptions are complex it is impossible to auto teach from all the components of the perception. It would be incredible to correlate anticipations with each pixel of an image. Like in the case of reactive agents intermediate categorisation of the perceptions is absolutely necessary to obtain perceptive invariants and to facilitate strategies. But unlike the reactive system, the hedonic agent has to enact emergent and evolving distinctions in moving environments. Through the history of coupling with its environment, it enacts a world of meanings which is pertinent for its pragmatic anticipations. Semantics comes from pragmatics.

A true capacity of interpretation emerges. The abductive capacity lies in this capacity to interpret some signs in the situation to perform good anticipations.

But all the auto teaching processes suppose an active experimentation of the different possible strategies. That supposes a huge number of effective trials in the environment. The revaluation of hedonic anticipations and strategies are very slow. Thus only a very slowly moving environment is acceptable at the hedonic level. We want see now how a much more quicker revaluation of hedonic anticipations and strategies are possible with eductive auto teaching.

2.3 Eductive level

The adjective "eductive" comes from the old English. It characterises the mind's capacity of an agent when it simulates its possible dynamics in its environment. Thus with eductive processes we have now to understand how an autonomous system can perform sensorimotor anticipations. The main idea is that a sensorimotor system can immediately correct a bad incremental sensorial anticipation because it sees during the next sensorimotor cycle what should have been the right anticipation. We shall use this crucial fact for grounding a cognitive paradigm of self-supervised sensorimotor learning.

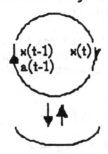

Sensorimotor cycles

Environment

fig 3. Auto teaching of sensorimotor anticipations : at each sensorimotor cycle the hedonic agent can correct what it has anticipated to see by what it sees now. Because of learning theory, the huge number of sensorimotor cycles allows to postulate quasi-bayesian sensorimotor anticipations.

Learning theory allows to postulate excellent sensorimotor anticipation capabilities during one sensorimotor cycle. Indeed, a sensorimotor system faces many cycles, each one providing an example of the space of possibilities : according to learning theory and the huge number of examples [Vapnik & Chernovenskis, 81] [Baum & Haussler, 87], it can reach quasi-bayesian anticipations of its own perception dynamics.

In a moving world, anticipation capabilities maintenance rely on the ability to learn faster than the relevant changes. In the following, this hypothesis is named "learnable moving world". Thus, under this hypothesis the following proposition is a potential property of any sensorimotor system.

Property (sensorimotor anticipation capacity)

a- In a learnable moving world, a sensorimotor system can acquire a quasi-bayesian estimation of its own dynamics $p(x'|x,a)$, i.e. the probability to perceives x' at the end of the sensorimotor cycle, if it perceive x at the beginning and performs a.

b- Together with its strategy function, this anticipation capacity also leads to a quasi-bayesian estimation of its own perception dynamics $p(x'|x)$, as a Markov chain. It can be iterated on a number of sensorimotor cycles, up to a certain horizon, depending on the quality of the bayesian estimation.

The above property is not a strong property, such as the one postulating a correspondence between mental and world states. No principle of correspondence between what is perceived and what happens "really" in the outside world is needed, i.e. no representationist hypothesis is necessary. The only useful hypothesis due to the recurrent nature of the cognitive system is (i) what it is perceiving inside can be correlated together (ii) what will just happen can be correlated with what has just happened. Here everything happens inside the same cognitive system, by auto teaching, in an autonomous way.

Thus metadynamics for hedonic and sensorimotor anticipations in the close future from the perceptions of the close past are possible to obtain better abductive capacities.

Because of the possibility of a hedonic system to auto teach sensorimotor anticipations, any hedonic system has the possibility to perform hedonic anticipations by an eductive self-reinforcement, for example by generating inside itself a huge number of possible virtual trajectories.

We can summarise this discussion by explicating the deep potential property of any auto teaching hedonic agent . All we need for the following is not a representationist hypothesis but the following one.

(the eductive autonomy hypothesis)

Any hedonic agent can auto teach its own perception dynamics and revise its hedonic anticipations and strategies in a close future by eductive self reinforcement, i.e. by using its sensorimotor anticipations and simulating its possible trajectories.

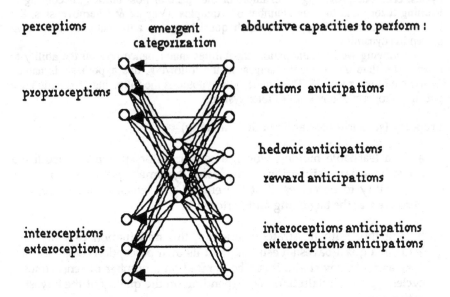

fig 4 : the eductive agent has a cognitive system which acquires abductive capacities to perform, from its perceptions, anticipations of the next sensorimotor state and its hedonic value. The cognitive system is a recurrent network (the back arrows) whose dynamics provides the anticipations and whose metadynamics allows to learn from all its internal events, i.e. auto teaching. The eductive auto teaching allows a quick revision of the sensorimotor anticipations and of recategorisation in order to re-evaluate the hedonic anticipations and strategies.

The first advantage for a hedonic agent of eductive self reinforcement is the possibility to replace the relatively limited number of examples provided by its unique real trajectory by a huge number of examples provided by a huge number of virtual trajectories.

The second advantage of eductive self reinforcement is the possibility to re-evaluate the hedonic anticipations when the sensorimotor dynamics changes because of changes in the environment.

These advantages are crucial in the case of a breakdown. Indeed a breakdown is an anticipated failure in the usual dynamics of the cognitive system : the anticipated action gives a very bad hedonic anticipations for diverse reasons such as changes in the environment. So by eductive auto teaching the agent can re-evaluate the hedonic anticipations and the hedonic strategies.

2.4 Construction of the   job's architecture and categorisation

A job is a simple process where the reward arrives at its completion time which can be a random variable. The concept is very important because most of the tasks present a delayed reward at its completion time. Some generalisation of the concept can be made, such as other intermediate rewards but the idea of a completion time remains and also that the essential of the reward is obtained at such time.

From the family of basic sensorimotor jobs, the hedonic agent can construct another job at a higher scale of time. Such construction has to take into account both the precedence constraints and the alternatives between the basic jobs. This kind of compositionality of jobs is very characteristic. And when a level of jobs is well trained and the strategies are very abductive by auto teaching, the construction can continue at a higher level of time or complication. So the architecture of jobs presents more and more layers, each layer beginning to be built only when the precedent is sufficiently achieved.

Such process is very close to the constructivism of Jean Piaget. We summarise the previous discussion in the following modelling hypothesis for a hedonic agent at a particular moment of its development.

Hypothesis (the job's architecture of the hedonic agent)

The hedonic agent disposes, through its activity of categorising situations, of a Random Arborescence of Jobs (RAJ) such as (i) the arborescence describes the precedence constraints between jobs (ii) new jobs can be added randomly to the arborescence at completion time of a job (iii) each job can be a RAJ at a smaller scale of time and this recursive operation can be reproduced until the family of basic sensorimotor jobs is reached.

The space of perceptions is generally very complex. We have already seen above the role of an intermediate categorisation in performing anticipations. We have now to discuss another fundamental role of categorisation on the job's architecture. This another role of categorisation is to pre select at each completion time and according to the situation the family of alternative jobs which have to be submitted to experience

But the hedonic agent has also another metadynamics to modify its current categorisation. By this way it can generate new distinctions and split a job into more specialised and adapted jobs. Thus it can modify the structure of its random of jobs in an adaptive way.

By compositionality and specialisation a huge number of jobs can emerge. For each job there is a necessity to do hedonic anticipations in order to enact abductive strategies. And a cognitive history is nothing else but a long alternated chain of categorised sensorimotor situations and chosen jobs. Because of its big length the choice inside the chain becomes a new and fantastic complex question for a hedonic agent with bounded cognitive capacities.

3 The hedonic agent as an open-ended auto teaching process

Now the question is how the hedonic agent can choose at each decision time the job it wants to submit to experience by eliminating all the other alternative jobs. This question of how its abductive capacity emerges becomes still more complex if the architecture of jobs is structured as a very large random arborescence as previously discussed.

The advantage of choosing a job is both the expected reward and a better knowledge about this job. There is a permanent compromise between choosing well known jobs with a better expected reward and less known jobs with a less expected reward : there is in fact a possibility that the less known one becomes better after experimentation. New questions emerge : when is it better to go on experimenting less known job ? or to cease experimenting for continuing with the much better known one ?

This compromise is the one of exploration/exploitation. It lies at the core of the adaptation processes as pointed out by [Holland, 75]. The theory of this compromise is those of bandit processes [e.g. Gittins, 89]. Because the concept of bandit process enlarges the one of job, it is very convenient to model the random arborescence of jobs by the same type of arborescence of bandit processes. By this way, the theory of bandit process provides a model of the perfect hedonic agent. After presenting this model the aim of this part is to envisage how an agent with bounded rationality can be modelled. What is obtained finally is a model of hedonic agent as an open-ended auto teaching process which works forwards in time.

3.1 The perfect hedonic agent and its forward induction policy

The perfect hedonic agent has, at each decision time, to make the choice which maximises its total expected discounted reward for an infinite horizon. Furthermore its decision process is not a unique process but a random arborescence of decision process with their precedence constraints.

The choice of a strategy would be of a fantastic complexity if it was treated by the backward induction policy which uses the familiar iterative construction backwards in time like in the dynamic programming. Even for a perfect hedonic agent (and a fortiori for a bounded rational agent of the next part) the a priori wish is of much more simpler policy.

The fundamental idea is then to search a forward induction policy which uses an iterative construction forwards in time. The simplest type of forward policy is an index policy : for each alternative job an index is evaluated and the alternative with the best index is chosen. Such index policy is very convenient if we think of the bounded cognitive capacities of an agent. But, if we want at least one index policy to be possible, it is shown that the form of discounting must be exponential. We can easily agree with this usual form of discounting and we always suppose in the following an exponential discounting. The question of big complexity is thus to find the policy inside jobs and between alternative jobs to maximise the total expected reward:

$$E\left[\sum_{i=0}^{\infty} \alpha^{t_i} r(x_i, a_i)\right]$$ where α is the discount rate and $r(x_i, a_i)$ the reward at time t_i

Surprisingly, in the case of a Random Arborescence of Bandit Processes, if the discounting is exponential, we don't loose any optimal policy by restricting to forward policy ; and we don't loose any optimal forward policy even by more restricting to index policy as shown by the following result [Gittins, 89] :

theorem : for a Random Arborescence of Bandit Processes, the class of index policies, the class of the forward induction policies are the same as the class of optimal policies.

The random arborescence has the same meaning as in the hypothesis of architecture of the hedonic agent (§I.4). At each completion time, the hedonic agent disposes of a family of alternative bandit processes among which only one has to become active while the others remain frozen. And a bandit process is simply a semi-Markov decision process completed by the frozen action.

Semi-Markov decision processes are a little more general than Markov decision processes. At each decision time of a Markov decision process, an action has to be chosen in a finite set $A(x)$ depending on the perception state x. The resulting state is x' with a probability $p(x'|x,a)$, the mean reward is $r(x,a)$[1] and the decision problem is to find a strategy $a=g(x)$ to maximise the expected total reward. In a semi-Markov decision process the next decision time is not a priori fixed as in a Markov decision process. It is indeed randomly distributed with some distribution $f(\Delta t \mid x, a, x')$.

Jobs are particular semi-Markov decision processes where the reward occurs essentially at completion time of the job, but the completion time is random. Because it includes both jobs and Markov decision processes, the class of semi-Markov process is a very large framework to modelize decision process as usually thought.

Let us turn now toward index policies for family of bandit processes. We have seen that there is no lack of rationality by choosing an index policy. The following result indicates that they are all equivalent to a basic index policy :

Theorem :

(i) Gittins index policy : the Gittins Index policy is an optimal policy for a family of alternative bandit processes. And every index policy is based necessary on an increasing function of the Gittins index

(ii) the Gittins index : A semi-Markov decision process (i.e. the origin of time is not a accumulation point of decision times) has a unique policy g which is deterministic, stationary and Markov and a stopping time τ which maximise the equivalent constant reward rate :

---

[1] The mean reward don't depend from x' because $r(x, a) = \sum p(x' \mid x, a) \, r'(x, a, x')$

$$v(x_0) = \underset{g,\tau}{Max} \frac{E \sum_{0 \le t_i < \tau} \alpha^{t_i} r(x_i, g(x_i))}{E \int_0^\tau \alpha^t dt}$$

(iii) the calibrating procedure : The Gittins index can be equivalently defined from the following dynamic programming equation where a *retirement option* M is added :

$$V(M,x) = Max\Big\{M, \underset{a}{Max}\ E[r(x,a) + \alpha V(M,x')]\Big\}$$

$$v(x) = Min\{M(1-\alpha)|V(M,x) = M\} = Sup\{M(1-\alpha)|V(M,x) > M\}$$

There is no loss of generality by choosing Gittins index policy. Such policy has a very intuitively meaning if we introduce the simplest bandit process : a *standard bandit process* is a process which provides a continuously constant reward equal to its index. In (ii) the bandit process is equivalent to a simple bandit process with the same index during the same time. In (iii) the retirement option M means that the bandit process continues playing until the expected reward is superior to M ; it stops in the contrary case and the retirement option is received. By construction, $V(M,x) \ge M$ always. When M is very big, the retirement option happens immediately and $V(M,x) = M$; when M is very small, the retirement option happens later or even never and $V(M,x) > M$. Between these two cases, there is a special M_0 where the retirement option happens still immediately but doesn't happen for inferior to M_0 values. But the retirement option M_0 provides the same reward as a standard bandit process with index $M_0(1-\alpha)$ from discrete time 0 to infinite horizon. Thus $M_0(1-\alpha)$ is the index of the standard bandit process exactly equivalent. We find back the same meaning as in (ii).

What we have finally to understand is the deep nature and theory of the compromise between exploitation of a well known bandit process and exploration of a less known one. In one of its interpretation of the calibrating procedure above in (iii), the theory of bandit processes belongs to the bayesian control adaptive theory.

Let us consider the simple example of the family of two alternative bandit processes $\{S(v), S'(1/2)\}$. The bandit S' provides the mean known reward of 1/2 and plays the role of a retirement option. In contrast the bandit S(v) provides 1 with the unknown probability v and 0 with the probability 1-v. Only what you know at first is the uniform distribution on [0,1] for the unknown probability. Suppose your discount rate is such that the length of the repeated game is quite long. Probably you avoid the strategy of immediately and always playing the known bandit S'. You will try the unknown one in the hope of a better index. Suppose after playing a few games that your estimation of v is a little less than 1/2. Intuitively it is clear than even if the estimation of v is a little less than 1/2 you feel the necessity to continue to experiment until the difference is sufficiently certain. But if you continue exploring the unknown bandit, when do you decide to play definitely either S or S' ?

In the bayesian framework - i.e. if the initial a priori distribution is known, here the uniform distribution on [0,1] - this problem of stopping or continuing to play the unknown bandit has a rigorous solution given by the calibrating procedure of the previous theorem : that depends if the index of the unknown bandit is superior or inferior to those of the unknown one, i.e. 1/2.

Let us remark that the state x of the bandit process in the calibrating procedure is now the parameters of the probability law to have some observation y (here the reward 1 or 0) if the action a is performed (use the unknown bandit or the well known one) : x represent the present knowledge about the bandit after previous experimentation. After performing the optimal action a, y is effectively observed and the probability law is revised with the Bayes rule. The metadynamics for the revision of knowledge is thus the Bayes rules.

But the very subtle point in evaluating the index is coming now : this evaluation takes into account all the virtual change of knowledge which can occurs in the future according to the initial probability law. Another very suggestive way to present this virtual process is as an eductive one : for a given policy of the bandit process, all happens if the evaluation is the mean of rewards until time τ of all virtual trajectories generated from the initial probability law with the bayesian rule as metadynamics for the revision of knowledge.

By this way, paradoxically, a less known bandit with less expected reward can have a better index than a better known bandit with a better expected reward. For example, according to the bandit theory, a bandit with 2 successes and 3 failures (mean reward : 0.4, index : 0.56) is better than a bandit with 40 successes and 34 failures (mean reward : 0.54, index : 0.55) for a discount rate of 0.95. Indeed the less known bandit has more possible futures than the best known one. Then better possible futures relatively to the best bandit can emerge after a few experiences and provide better rewards for the long term.

What we have understood now is the optimal cognitive strategy in the bayesian framework to achieve the compromise between exploration and exploitation in a family of alternative bandit processes. This cognitive strategy takes into account both the effective reward and the gain of knowledge. Its corresponding metadynamics is the bayesian revision.

Hypothesis (the perfect eductive hedonic agent) :

> Let us consider a hedonic agent defined by its random arborescence of bandit process. The perfect hedonic agent (i) uses an index policy to choose at each completion time the next possible bandit process which has the best index (i.e. whose evaluation is equivalent to the perfect eductive process of generating all the possible trajectories in the future) (ii) uses inside the bandit process the optimal strategy corresponding to the index until the next completion time (iii) revises its knowledge with the Bayes rule.

3.2 The hedonic agent with bounded rationality

We have now a model of the perfect hedonic agent. What is to be kept is its two main features (i) the forward character of the index policy with its remarkable tractable computationality (ii) a good compromise between exploration and exploitation, i.e. a good cognitive strategy to take into account both the effective reward and the gain of knowledge.

But this perfect model is untractable despite the forward character of the policy : except in the simplest cases[2], the computational cost of calculating indices is cumbersome. What we want is an abductive capacity to estimate in the flow of time indices and strategies. We have thus to search approximated but more tractable metadynamics for these evaluations. For this aim, we have to take inspiration both in the above theory and in the connexionist area.

The general movement of ideas in the connexionist area are very close to the bayesian point of view of the above theory. Every supervised learning which tends to minimise the quadratic error provides a quasi-bayesian estimate. This quasi-bayesian estimate is obtained by a metadynamics which change the weights of the network. The knowledge is distributed on these weights. Because every learning process of the hedonic agent is an auto teaching process, it is also a self-supervised learning process. Let us also notice the general method TD(λ) of Temporal Difference to improve the anticipations by taking into account the recent past with an exponential decreasing importance depending on the parameter λ [Sutton, 88].

Before we consider the evaluation of the indices, it is important to remember the movement of idea in the reinforcement learning area. In this area the standard method is the Q-Learning rule associated with the Boltzmann rule for the choice of actions [Watkins, 89].

Definitions :

(The Q-learning rule) : The Q-Learning rule is a reinforcement metadynamics which tends to minimise the error with the functional equation of dynamic programming for the hedonic anticipation Q(a, x) of an action a in a perceived situation x providing an immediate reward r(x, a) and a new situation x' of value V(x').

$$\Delta Q(x,a) = \varepsilon \left[r(x,a) + V(x') - Q(X,a) \right]$$
$$with \ V(x') = \underset{a}{Max} \, Q(x',a)$$

(The Boltzmann rule) : The Boltzmann rule is a probabilistic rule where the actions are chosen proportionally to the exponential of their expected hedonic value Q(a, x) :

$$p(a|x) \propto e^{Q(a,x)}$$

For evolutionary processes, Holland proposes a justification of an exponential offspring for the best individuals [Holland, 75]. His justification is also based on the

---

[2] The reader can see many remarkable examples in the book of Gittins where the evaluation of indexes is tractable. The special case where analytic expression can be found for index is considered also as "simple". For example, the evaluations of job indices are quite simple.

theory of bandit processes and the compromise between exploration and exploitation. More study is to be done to extend this justification to the present self reinforcement learning framework.

We have now to be more precise on the metadynamics for the evaluation of the indices. The Q-Learning concerns only one decision process. If we have many decision processes, each is submitted to a pressure of selection from the others. And we cease to be interested in the optimal strategies in the long term for a decision process. We are interested in the strategies which have a good return for a shorter term. And it is precisely the role of the theory of bandit processes to provide the best stopping time and the best strategy for this best term. In the case of a random arborescence of bandit processes, we have thus to modify the Q-Learning rule by introducing the retirement option M (cf the calibrating procedure of the above theorem). I propose to name this new rule the I-Learning rule because the role of this rule is to calculate both the anticipated quality of actions and the indices.

definition : (the Index-Learning rule or I-Learning rule)

The I-Learning rule is a reinforcement metadynamics which tends to minimise the error with the functional equation of calibrating procedure for the hedonic anticipation Q(a, x, M). The main change is that the retirement option M is applied if the value of the new situation x' is inferior to M.

$$\Delta Q(x,a,M) = \varepsilon \left[r(x,a) + V(x',M) - Q(X,a,M) \right]$$
$$with \ V(x',M) = Max\{M, \underset{a}{Max} Q(x',a,M)\}$$

Let us consider the cognitive strategy of a simple hedonic agent without eductive capacities. The best it can do is to re-evaluate at each time the anticipated quality of actions for a close future by using what has just happened in the close past : that is the role of metadynamics such as I-Learning rule with TD(λ). It has to choose its jobs according to their indices and its actions inside its jobs according to their anticipated quality of actions.

But the hedonic agent has also another metadynamics : it can modify its current categorisation. By this way new distinctions are generated and it can split a job into more specialised and adapted jobs . Thus it can modify the structure of its random arborescence of jobs in a adaptive way. The most important role of the neural network is to perform parallel distributed processing. What is more important is probably not an exact evaluation of the index but a large scanning between a huge number of jobs. The categorisation metadynamics is the most important mean for a simple hedonic agent to change its main jobs. The categorisation derives both from the basic distinction capabilities of the agent and from the need to better hedonic anticipations. The above discussion about the metadynamics of a simple hedonic agent allows to reformulate the simple hedonic level as follow :

Hypothesis (the simple hedonic level) :

the hedonic agent has continuously to manage (i) the change of its random arborescence of generalised jobs by its recategorisation metadynamics (ii) the revision of hedonic anticipations, of jobs indices and of action strategies through metadynamics (like I-learning rule and Bolzmann rule) which have to be implicitly near optimal in respect to the perfect hedonic agent.

More work has to be done on the character more or less optimal of metadynamics like the I-Learning rule or of the Bolzmann rule in respect to the compromise exploration/exploitation over a long term horizon. The long term horizon comes from the slow character of the metadynamics.

The cognitive level attained by the simple hedonic agent correspond in neuroscience to the primary consciousness as stated by Edelman. But this level doesn't allow to change rapidly either categorisation or hedonic anticipations, jobs indices and actions strategies. For that we have to shift to the eductive hedonic level.

Let us consider finally what is the cognitive strategy of the eductive hedonic agent. The eductive agent, because it has an estimation of its own perception dynamics (cf. §I.3), has the possibility to generate a big number of virtual trajectories. By this way it can re-evaluate at each time its indices and strategies in the same way as the perfect hedonic agent but only approximately : the approximation to the best index and strategies of actions depends only on the time it has to perform its revision ; the more trajectories it can generate, the more accurate its evaluation is. In fact revaluation is still a job and the present theory can say when this special job of revision of knowledge has to be stopped. Let us note that the eductive mechanism is nothing else but auto teaching of a simple hedonic agent with its particular dynamics and metadynamics by travelling in many possible futures compatible with its present knowledge. And the perfect hedonic agent is nothing else but a simple hedonic agent which can travel in all its possible futures.

Naturally it doesn't re-evaluate at each time its indices and action strategy : if its perceived situation is well known it can take the current strategy and revise slowly with the I-Learning rule. In this case its current abductive capacity works very well. But sometimes, it can verify on its own perception dynamics (i) that a perceived situation is relatively unknown or (ii) that its perception dynamics is moving or still (iii) that there is a breakdown in the close future. In such case, it has to re-evaluate its indices and strategies. By this simple way new global strategies and categorisations can eductively emerge by simple recombination of jobs in its arborescence of jobs. New abductive capacities emerge.

Hypothesis (eductive hedonic level) :

the eductive hedonic agent is a simple hedonic agent with an imperfect knowledge of its own dynamics. At any time it disposes of a supplementary job which consists by travelling in many possible futures compatible with its present imperfect knowledge in revising (i) its current categorisation (ii) its current hedonic anticipations. This knowledge revision job is an explicit management of its compromise between exploration and exploitation and obeys to the common rules of job : it occurs when it has a sufficient index (i.e. in the case of a breakdown of different types) and has to be more or less completed in respect to a stopping time (for an infinite stopping time, the eductive agent is the perfect hedonic agent).

4 Conclusion

What was considered here is the class of agents which have to maintain their viability in various and changing environment despite their bounded cognitive capacities. One of the simplest way to think such an agent is as a hedonic agent with its forward induction policy and its global open-ended process : at each time, the history of its coupling with the environment is synthesised in its current state of knowledge resulting of its auto teaching metadynamics ; and at each time it has a more or less large window around the present, where its dynamics uses the recent past to do anticipations for the close future ; at each time its metadynamics revises the current state of knowledge by taking into account implicitly or explicitly the compromise between exploration and exploitation which is essential to various and changing environment. Then the window shifts to the next decision time, in a recurrent way for this open-ended process.

A forward induction policy is absolutely essential for an agent with bounded cognitive capacities. The evaluation of indices for job is first to provide a forward induction policy. This evaluation is still performed in a forward way. We have here a teleonomic figure where all happens effectively by dynamics forwards in time but also everything happens as if the hedonic agent follows some objective. That is a remarkable point. This teleonomic point of view goes beyond the behaviourism. The hedonic agent is effectively conditioned by its environment ; but conversely it acts to its environment in order to transform it in an ecological niche for it.

No auto teaching can happen without a hedonic principle. That defines the hedonic level hypothesis. What we have at this level is a theory of emergence of how-know in bounded rational agents : the hedonic agent enacts a useful categorisation for its hedonic and sensorimotor anticipations, its sensorimotor knowledge, its random job's architecture. This cognitive level comes before the symbolic level. We meet here the point of view of Piaget, who claimed that the symbolic knowledge of a child comes after a prealably acquired sensorimotor knowledge.

We have also a theory of emergence of abductive capacities. Peirce states abduction as "the mind's capacity to guess the hypothesis with which experience must be confronted, leaving aside the vast majority of possible hypotheses without examination". The role of experience becomes essential. The next crucial question for hedonic agents is the lack of experience. This question has to be treated in a society of such agents.

References

Aubin J.P.,1991. Viability Theory, Birkhäuser.

Baum Eric B., David Haussler, 1989, What Size Net Gives Valid Generalization ? Neural Computation 1, 151-160 (1989).

Bourgine P., F. Varela 1992. Towards a practice of autonomous system. in Towards a practice of autonomous system, F.Varela & P.Bourgine (ed). MIT Press/Bradford Books.pp 3-10.

Bourgine P., 1993, Viability and pleasure satisfaction principle of autonomous systems, in Imagina-93 proc.

Brooks R., 1991. Intelligence without reason. IJCAI-91,Sydney.

Brooks R., 1991. Intelligence without representation. Artificial Intelligence 47, Jan., 139-159.

Gittins J.C., 1989, Multi-armed Bandit. Allocation Indices, John Wiley & Sons

Edelman, G, 1992, Bright Air. Brillant Fire : On the Matter of Mind, Basic Books.

Holland, J.H., 1975. Adaptation in natural and artificial systems. Ann Arbor : the university of Michigan Press.

Kohonen T., 1984. Self-Organization and Associative Memory. Springer Verlag.

Langton C., 1989 . (ed) Artificial Life I, Addison Wesley.

Langton C., 1992,Life at the edge of chaos, in Artificial Life II, Addison-Wesley, p.41-92, 1992.

Meyer Jean-Arcady, Wilson Stewart W., 1991, From animals to animats, M.I.T./Bradford Book, Cambridge,MA.

Nicolis G., I.Prigogine, Exploring Complexity: An Introduction, R.Piper GmbH & Co. KG Verlag, 1989.

Peirce Charles S., Textes fondamentaux de sémiotique, Méridiens Klincksiek, Paris, 1987.

Petitot J., 1990, Physique du sens, editions du CNRS.

Rosh E., 1978, Principles of Categorization, in Cognition and Categorization, ed. E.Rosh and B.B.Lloyd, Lawrence Erlbaum, Hillsdalle, N.J., 27-48.

Rumelhart D.E. and J.Mc Clelland, 1986, Parallel Distributed Processing, MIT Press/ Bradford Books.

Simon H.A. (1976) From subtantive to procedural rationality. Method and Appraisal in Economics, Latsis S.J.(ed.), p. 129-148. Cambridge University Press, Cambridge.

Sutton, R.S., 1988, Learning to predict by the methods of temporal difference. Machine Learning., 3, 9-44.

Valiant L.G., 1984, A theory of the learnable, Communications of the ACM V27, n°11 pp. 1184-1142.

Vapnik V.N. et Y. Chervonenkis, 1981. On the uniform convergence of relative frequencies of events to their probabilities. In Theory of probability and its applications, XXVI, pp 532-553.

Varela F., 1979. Principles of Biological Autonomy, North Holland, Amsterdam.

Varela F., 1986. Trends in Cognitive Science and Technology.in:J.L.Roos (ed.), Economics and Artificial Intelligence, Pergamon Press, Oxford, pp. 1-8.

Varela F., E. Thompson & E. Rosch, 1991, The Embodied Mind, MIT Press.

Varela F., P.Bourgine, 1992, Towards a practice of autonomous system, MIT Press/Bradford Books.

Walliser B., 1993, A spectrum of cognitive processes in game theory, in Second European Congress on System Science, Prague, oct 93.

Watkins C., 1989, Learning with Delayed Reward, PhD, Cambridge University Psychology Department.

The Conceptual Framework of MAI²L*

Donald Steiner[1], Alastair Burt[2], Michael Kolb[2] and Christelle Lerin[3]

[1] Siemens AG, c/o DFKI, P.O. Box 2080, D-67608 Kaiserslautern, Germany
[2] DFKI, P.O. Box 2080, D-67608 Kaiserslautern, Germany
[3] Steria, 12 Rue Paul Dautier, F-78140 Velizy-Villacoublay, France

Abstract. This paper describes the key concepts behind the Multi-Agent Implementation and Interaction Language, MAI²L. The three major design requirements of rationality, reactivity and generic cooperation are explained and the architecture to support them is detailed.

1 Introduction

The Multi-Agent Implementation and Interaction Language —MAI²L— was developed in the Esprit Project 5362, Integrated Multi-AGent INteractive Environment, IMAGINE. It is used for building systems composed of autonomous, cooperative agents. In describing the concepts we shall answer the following questions from IMAGINE's point of view:

1. What is a situated, rational agent?
2. What makes it cooperative?
3. What are the mechanisms that are required of MAI²L to support systems of such agents?

The next section describes the general design principles; it is followed by an explanation of how, first, the internal agent architecture and, second, the inter agent cooperation mechanisms meet these principles; we then describe how such a system might be used and implemented.

This report is an abbreviated, simplified account of [7]. The operational and implementational details of MAI²L are discussed in [6]. The tone we use in the rest of this paper is informal; [3] defines the key aspects of MAI²L more rigourously in terms of abduction and shows how MAI²L cooperation protocols may be generated by abductive inference procedures.

1.1 The Underlying Concepts

The overall aim in IMAGINE is to have a single framework that captures cooperation between humans in the workplace as well as interaction between software processes. In this way we build on the results of research in Distributed Artificial Intelligence for the software side and Computer Supported Cooperative Work

---

* This work was funded in part by the Esprit II Project 5362, IMAGINE.

for the human side. An agent in this general framework may represent software, a human or a combination of both.

An IMAGINE agent has three important characteristics: rationality, cooperativity and reactivity. We therefore adhere to three design requirements that must be met in the multi-agent framework described here and in the language and applications derived from it. The first is immediately derivable from our assumptions about the nature of agents:

Requirement 1 Rational Agent. *An agent should structure its behaviour in a way that, as it reasons, will optimally satisfy its goals. An exact definition of optimality is dependent upon the type of goals of the agent and its ability to reason about achieving them.*

Thus, an agent will carry out a task, only if it thinks it will lead towards a goal. If it thinks a task carried out at time T leads optimally to its goals, it will indeed carry out the task at time T. The rational agent requirement is desirable because it describes an agent that is both flexible and predictable. The agent is flexible because there may be more than one way to reach a goal. It selects the alternative that is appropriate to the environment at a given time. The agent is predictable in that, knowing enough about the goals and beliefs of the agent, we can work out which alternative way of satisfying its goals it will choose. Problems arise in applying this requirement when:

1. it is not clear which of several goals the agent should work towards; and
2. the agent has less than perfect knowledge about the state of the world, present and future, or about the preconditions and effects of tasks.

The reasons why a machine agent may possess less than perfect knowledge are:

1. incorrect information in its knowledge base;
2. an incomplete knowledge base; or
3. insufficient resources to draw inferences from the knowledge base.

Because of such constraints, we refer to an agent's rationality as being *bounded*.

Agents in a multi-agent system may find that the optimal means (from their point of view) to reach their goals is to get other agents to carry out certain actions. The process of several agents working out a future course of action together and then carrying it out is how we define cooperation. Here, we make use of a second requirement:

Requirement 2 Generic Cooperation. *When several agents cooperate, they should do so in ways that are, in important respects, independent of a particular domain.*

Generic cooperation arises from the fact that the agents are rational. Our notion of cooperation primitives and cooperation methods derives from the second requirement. The general form of the cooperation methods is independent of the domain, although the particular goals, plans and tasks that are mentioned in the methods is, of course, domain dependent. The assumption that there can be general purpose cooperation is related to the assumption, common in AI, that there are general purpose planning techniques.

The application domains in which IMAGINE agents operate impose certain real time constraints on an agent, and so we have a third requirement:

Requirement 3 Reactivity. *The architecture of an agent should be such that it can react in a timely fashion to changes in the environment.*

The exact definition of "timely" will vary, but certain general mechanisms are used to support this requirement: describing certain key events in the world and the appropriate reaction to them; constant monitoring of the environment to detect the key events and trigger the appropriate reaction; and scheduling of resources so that enough are always at hand to meet contingencies.

The three requirements are not orthogonal; there are dependencies between them:

1. **Rationality and Generic Cooperation.** Rational agents will consider the actions of other agents when trying to reach their goals; this is how cooperation arises. Cooperation is generic precisely because rational agents share certain characteristics. They all must be able to reason about goals and the adequacy, quality and efficiency of plans.

2. **Generic Cooperation and Reactivity** The cooperation an agent engages in is dependent on the role it plays. This role is triggered in a reactive way through situation assessment. The commitments that an agent undertakes during cooperation to carry out actions in the future must allow the agent the flexibility to meet the reactivity requirement.

3. **Rationality and Reactivity** Where an agent can reason that the optimal course of behaviour is for it to react quickly to the environment, then to be rational is to be reactive. Meeting the reactive requirement may mean that the agent is not able to devote as many resources as it would otherwise do to find the optimal plan. Reactivity puts a bound on rationality.

2 The Behaviour of Goal-Directed Agents

A crude outline of the architecture of an Imagine multi-agent system is given in Fig. 1. A more detailed description of this architecture is given in [10]. Domain specific capabilities of an agent are encapsulated in the *body* of the agent. The *head* reasons about the functions of the agents body and controls the body's execution. In a multi-agent system the heads of agents are responsible for informing one another about their body's capabilities, for making *commitments* about their body's future actions and ensuring these commitments are met. The *communicator* implements a message passing service between agents' heads. We will not mention it further here, and direct the interested reader to [11, 2]. We likewise ignore the domain dependent details of the agent body. This leaves the agents' heads, which we examine in this section, and their cooperative interaction, which we examine in the next.

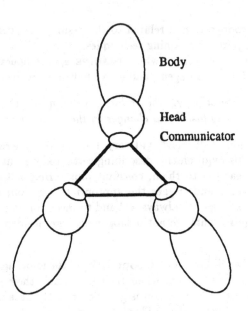

Fig. 1. The Architecture of an Imagine Multi-Agent System

The internal architecture of an agent head is depicted in Fig. 2. The upper nodes represent knowledge processing elements, the lower nodes represent elements in the knowledge base. Conceptually the four knowledge processing elements operate concurrently and all of them are defined through plans. This allows the possibility that the agent may plan how to plan. We will examine each of the processing elements in turn.

2.1 Goal Activation

A goal is a description of a future state of the world. A rational agent is one that has choices to make about the actions it may carry out and chooses those actions that will bring about its goals. In the simplest model an agent comes into existence with one goal; it derives a course of action, that is a plan, to achieve that goal; it executes the plan; it terminates.

A more general model needs to have a more complex taxonomy of goals. Thus an agent may have more than one goal and these goals may be:

1. **Conjunctive** The agent may strive towards several conjunctive goals, aiming to achieve all of them, possibly at the same time, possibly at dispersed intervals in the future. There may be conflicts between the conjunctive goals, in particular they may compete for resources. The designer of a software agent should ensure that such no goal is continually blocked by others, and build mechanisms into the planning component that resolve resource conflict. The agent may have preferences between conjunctive goals, being prepared to devote more resources to some goals at the expense of others.

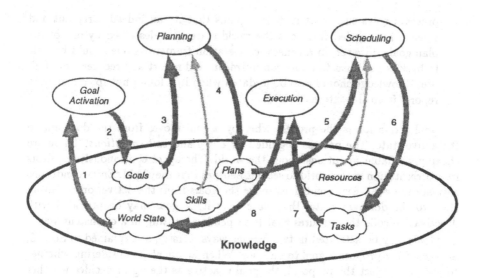

Fig. 2. Knowledge Processing Within an Agent

2. **Negative** The goal of an agent may be expressed in terms of properties that are not to hold in the future.
3. **Complex Temporally** Depending upon the agents ability to reason about time, its goals may be expressed in terms such as:
   (a) a property should always hold
   (b) a property should never hold, or
   (c) a property should hold at certain well defined time points or intervals.
4. **Subgoals** A goal may describe a subset of states of another. The former is then a subgoal of the latter.
5. **Predecessor Goal** Whilst planning an agent may derive a plan to go from a state in the future to a goal. The state may be referred to as predecessor goal to the ultimate goal.

In order to support the reactivity requirement in agents, the MAI²L model further differentiates goals by allowing them to progress through various states:

1. **Dormant** When a goal is dormant, the agent devotes no resources to it. That is the agent neither plans towards the goal, nor commits itself to a plan that would achieve it. The goal thus has the absolute minimum preference associated with it.
2. **Active** A goal is active when an agent is either planning towards the goal or has committed itself to a plan that it believes will lead to the goal.
3. **Executing** The goal is executing when the agent is carrying out a plan that it has reasoned will satisfy its goal.
4. **Achieved** A goal is achieved when the agent terminates the plan associated with it. We usually assume the agent's knowledge representation and

processing are such that it derives plans that it can indeed carry out and when it has carried them out the world state is as described by one of the plan goal. However, in a general cooperation framework one should be able to back out of unsustainable commitments and report and recover from failure. Knowledge should also be updated when it is found not to concur with reports from the external world.

Goal activation is the process whereby a goal moves from the dormant to the active state. The process is guided by *entry* and *exit conditions*. These are cheaply verifiable conditions about the world. The agent continuously monitors its representation of the world to see if it believes that they hold. Entry conditions activate the goal. Exit conditions move the goals from the active, or executing state to the dormant state: they *deactivate* it. A goal may be provided with deactivation code that ensures that it suspends executing in a consistent way.

Goals may be activated in two further ways. Firstly, as explained in Sect. 6, an agent may propose a goal to another. When an agent is considering whether to accept or reject the proposal, the goal is active as the agent decides whether the proposed goal is a suitable predecessor to one of its already fixed goals.

Secondly, user agents act as a user's stand-in in a multi-agent system. A user may arbitrarily rearrange the goal structure of his user agent in order to get it to represent his views correctly.

3 Planning

Once a goal is activated the planning component of the agent head may set about deriving a plan to achieve it. Planning is an intrinsic capability of a rational agent but the demands of reactivity may be a severe constraint on the amount of resources the agent can devote to deliberating its future actions. In some cases planning will be limited to minor modifications of predefined plans that have been carefully crafted by the agent designer to meet the situations that the agent is likely to encounter.

A MAI^{2}L plan consists of three components:

1. the *plan procedure*, that describes how a plan is made up of subplans,
2. the *preconditions* of the plan, which describe the state of the world immediately before execution, and
3. the *effects* of the plan, which describe the state of the world immediately after execution.

Because a plan is such a central notion in an IMAGINE multi-agent system, it is useful to distinguish various types of plan. We shall examine these types along three dimensions: how complex the compositional constructs of the plan procedure are, how runnable the plan is, and whether one or several agents are involved.

Complexity of Plan Procedures. **Plan** procedures describe how a course of action is composed of *subplans*. An agent can carry out certain subplans without further planning or interpretation: we call these *atomic plans*. Control constructs combine subplans to form plans. Simple planners can only reason about sequencing as the control constructs. However, in MAI²L all algorithmic components of our agent architecture, including cooperation methods, are captured in the plan formalism. This means that the control constructs are as powerful as those in a general purpose programming language, and include, not only sequencing, but also conditional branching, and iteration. The control constructs are exactly those that can be interpreted by the executor in the agent head. We also have the power of procedural abstraction in our plans.

Plan Completeness. Ultimately, the objective of planning is to construct a plan procedure that is executable. Initially, the planner may start off with a plan that has no procedure, only preconditions (the initial state of the planning problem) and effects (the goal state of the planning problem). We list below the types of plan that lie in between these two extremes and what operations need to be performed on them to generate an executable plan:

1. **Partial Plans** In general planning consists of finding actions to get from an initial to a goal state. In partial plans not all the actions have yet been specified. There remain *plan gaps*, consisting of further initial and goal states. Only when these gaps have been fully planned, is a complete description of the appropriate actions available. By making partial plans first class citizens in our planning formalism we allow for the interleaving of planning and execution, and for the passing of partial plans to another agent to involve it in the planning process.

2. **Abstract Plans** The components of a plan are state descriptors and subplans. These may be only abstractions of other states and subplans. In the course of deriving an executable plan, the planner will have to find concrete instances for the abstractions. In our plan formalism, where we represent general programs, subplan abstraction is equivalent to procedural abstraction in a programming language. A simple form of abstraction is the use of variables. A plan containing variables is a *plan schema*.

3. **Executable Plans** A plan in which all plan gaps have been filled and all states and subplans have been reduced to their lowest level of abstraction is executable; agents can run the plan procedure without further planning.

Single and Multi-Agent Plans. Plans may describe a course of action in which more than one agent takes part. The plan procedures, therefore, include *multi-agent plans*, where the agents which carry out the subplans are recorded. As the planning and scheduling process proceeds, a multi-agent plan is reduced to a single agent plan which is runnable by the executor of a single agent.

Commitment to Plans. An important feature of cooperative planning is *commitment* to plans: that is the degree certainty that an agent will perform a plan.

224

The defining characteristic of an agent is that it must choose between several courses of action. Before the agent makes any choice it is not committed to any of the alternatives. As the agent investigates the search space of plans it will narrow down its options. The demands of cooperation will naturally restrict an agent's freedom of choice. Exactly how its freedom is limited is determined by the communally agreed semantics of a particular cooperation environment. For example, in when people schedule appointments, there is a notion that an appointment is only "pencilled in". This means the person is not fully committed. Must a possible attendee leave that slot free until the date is confirmed? How long is he prepared to wait for confirmation? Answers to these questions may vary from context to context.

4 Scheduling

The plans of an agent are not immediately runnable. The ordering of the execution of subplans is not fully specified, and some simple tailoring to the internal resource availability is still required. This is the job of the scheduler. Where a task is to carried out by, or in coordination with, another agent the relevant synchronisation messages are generated by the scheduler. The scheduler also checks the triggering of conditions of plans. Its role in this respect is like that of the goal activation component.

5 Execution

The execution component routes messages to and from the communicator, calls body functions via the head-body interface, calls head functions via a meta call facility in the MAI^{2}L implementation language.

A head-body interface is, in fact, an aggregation of fine-grained interfaces which establish a link between the head and the collection of body elements composing a body. A body element provides one functionality which either can be invoked on its own by the head or which periodically provides data which are relevant for the agent's head.

The functionality provided by a body element may :

1. Be modelled in the agent's head by a body skill, and be invoked when the time for executing a task concerning this skill has come. In this case, the head-body element interface must provide the link head-body element, as a call to an external procedure, and the link body element-head, for the return of results which update the agent knowledge (in the agent's head).
2. Or happen on a continuous basis without any invocation from the agent's head (except a trigger order from the agent's head to make the body element begin its work); this is a perception functionality offered by a body element and acts as a basic component for the agent's situation assessment component. In this case, the head-body interface must update the right elements of the agent's knowledge (in the agent's head).

The functions of the body are embodied by procedures of plans. Thus, the head can fully plan about body functions. The MAI²L programmer needs only supply the input and output parameters, preconditions and effects.

6 Cooperation among Agents

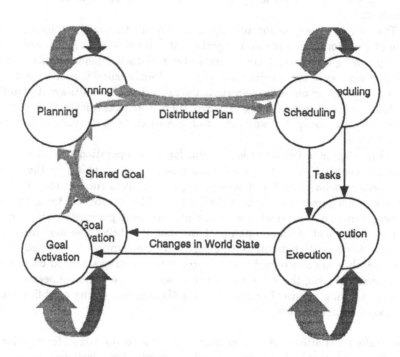

Fig. 3. Process View of Several Distributed Agents

Cooperation arises as several rational agents plan and execute their actions in a coordinated way. When an agent cannot generate and run plans on its own, its is obliged to communicate with other agents for information and to obtain commitments from the others that certain actions will be carried out or certain goals realised (see Fig. 3). In a system of agents a tangled web of interlocking commitments and information exchange will commonly arise. However complicated the cooperation becomes it will still share certain simple features that arise from the plan generating internals that are common to all agents. This is the basis for the notion of cooperation methods and primitives and is the way we meet our generic cooperation requirement.

Cooperation Methods. Cooperation methods are multi-agent plans used to construct and execute domain-specific multi-agent plans. As a consequence, they concern goals, plans and tasks of the agents involved. The knowledge that the agents have of each other determines the topic of the cooperation method. As an example, if an agent has an active goal it can not itself handle and it doesn't know about the skills of other agents, it must invoke cooperation about the goal. If it has knowledge about which agents could satisfy the goal, it will initiate cooperation about the plan for achieving the goal. Further, if a domain-specific plan is already committed to by agents, they will cooperate about the tasks to be executed.

The aim of a cooperation method is to commit the involved agents to the topic of the cooperation method. Agents that cooperate about goals try to delegate planning activities. A cooperation about plans is intended to fix a future course of actions of the involved agents, i.e. a domain-specific multi-agent plan. Agents cooperating on tasks coordinate the execution of a multi-agent plan they previously agreed upon. During task execution, a general information exchange is necessary to propagate results of a task execution or to deliver parameters to other agents.

A high degree of flexibility is essential for the cooperation methods. For instance, a *contract net* is a flexible cooperation method which, in the general case, neither determines the number of agents involved (in fact the number is unlimited) nor prescribes which of the agents will be members of the group. Obviously, predefined cooperation methods must be freely parameterisable. To simplify establishment of cooperation patterns that are likely to be used frequently, predefined cooperation methods have partly or even completely specified parameters, i.e. they are specialised plans. They allow invoking standard cooperation structures involving the same set of agents and/or the same cooperative links thereby reducing the usual preliminary negotiations concerning establishment of the cooperation structure.

Cooperation Primitives. A cooperation method involves a transfer of information from one agent to one or more other agents. But, furthermore, there is a special intention behind every information transfer. Both aspects are covered by a set of *cooperation primitives*. Cooperation primitives are basic agent head functions, describing communication among agents with a specific intention. Therefore, they are represented as plans, whose preconditions and effects fix the semantics/intention of the primitives and whose plans consist of a call to the head-function handling the communication (head-communicator-interface).

A cooperation primitive always takes a cooperation object and a set of agents (the receivers) as inputs. The cooperation object is either a goal, a plan, a task or unspecific information such as results, parameters, or other knowledge. The set of cooperation primitives and their intended use is given by the following list:

propose A proposal starts or continues a discussion among agents about a cooperation object. The chunk of information transferred by a proposal to other agents is in some sense hypothetical as the agents sharing this information have not yet committed to it.

refine	An agent sends a refinement if it can not commit to an object as it was previously proposed, but its evaluation of the cooperation object has led to an instantiation of the proposal, which it sends to other agents.
modify	A modification is similar to a refinement, except that the cooperation object sent by a modify differs in more than just an instantiation of the previous proposal. Refine and modify represent counter-proposals, but do not commit the agents.
accept	An agent indicates its acceptance of a cooperation object to other agents. Both sides, the sender and the recipients commit to the object.
reject	A rejection is used by an agent if it cannot commit to the proposed cooperation object. Accept and reject terminate a discussion about a cooperation object either successfully or with a failure.
order	An agent ordering a cooperation object imposes the acceptance of the object on the recipients. But, the recipients might still reject the object[4]. An order is only applicable if either the object has been previously discussed or an appropriate authority link between sender and receivers is established.
request	An agent can request particular information about a goal, plan or task from other agents.
tell	Tell is the answer to a previous information request.

The semantics of the cooperation primitives is not only given by their default meaning, but also depends on the cooperation methods that bundles a set of primitives. But every agent understands the cooperation primitives and might thereby be able to follow a plan describing a cooperation method it doesn't share.

In order to provide the flexibility required by human participants in a cooperative process, we allow for dynamic definition of new cooperation primitives. For example, one could introduce message types such as SUPPORT and OPPOSE in order to more efficiently communicate about intentions.

7 Relation to Other Work

MAI^{2}L is rather broad in its scope and consequently touches on several areas of research:

1. **Modelling Communication: Speech Acts** There has been much work recently trying to formalise the patterns of communication of people as they

---

[4] For example, if the object is a task, an agent may reject the order if it can not satisfy the task's preconditions. If, on the other hand, a task failed during execution, the reply to the order would be the result of execution, i.e. task-failed.

interact [12], much of it based on the philosophical notion of speech acts [[8]]. The distinguishing features of our cooperation methods are:

(a) they have a different semantic basis,

(b) they are highly flexible and integrated with the planning framework, and

(c) they encompass machine-human, machine-machine, and human-human interaction within a single system.

2. **Planning** [4] and [1] among others have proposed ways to reason about action that is as general as our model, basing their ideas on program logic. In MAI^{2}L we have carried this work to its logical conclusion by representing all algorithmic aspects of the system in a plan formalism.

3. **Situated Automata** [5] have a notion of agent that stresses the agent's ability to react quickly to changing situations. Our model is like theirs in that it uses goals to direct the "firing" of actions but our model also uses the notion of goals to direct a more deliberative planning process.

4. **Agentification of Software** There are several strands of work that attempt to "agentify" existing software, [9]: that is, give existing software explicit goals, allows them to operate in a distributed, heterogeneous environment. We share these aims. MAI^{2}L differs from these in that the current implementation is based on an inherently parallel language. This also aids in the conceptualisation of the problem.

8 Conclusion

We summarise here the computational mechanisms through which we are implementing the concepts.

Goal Activation. Goals can be activated either:

1. internally
   (a) Situation assessment requires real-time forward reasoning to allow for immediate reaction to changes in the world state.
   (b) The human must be able to activate goals.
2. Reaction to cooperation messages from other agents.

Agent Planning Capabilities. Agents are provided with planning capabilities which allow them to build plans for a given goal. The sophistication of an agent's planning capability may vary from agent to agent – reactive agents may have predefined plans linking their goals immediately to the corresponding tasks. MAI^{2}L is supported by a run-time planning engine, which can be used to automatically generate plans from goals. However, other planners can be used for an agent – they need only conform to the MAI^{2}L representation of goals, plans and tasks.

The planning capabilities must include the following inference mechanisms:

1. Finding plans for particular goals. Realised by effect matching – requires backward reasoning from the desired world state to an initial situation (e.g. the current world state).

2. Finding plans suitable for particular situations. Realised by precondition matching – requires forward reasoning from the current world state to the desired state.

3. Finding plans hypothetically within a "simulated" world. That requires a projector which is able to simulate the plan execution within a context and derive the expectable outcomes (situations that lead to the activation of new goals, hazardous situations, etc.).

Scheduling. When an agent reasons about which plan to use to achieve a goal, it does so hypothetically. When the best plan is found it must be moved from the hypothetical to the real world. This forms the semantic basis for the notion of commitment. At this point the corresponding task can be constructed and scheduled (task-specific attributes instantiated). A variety of mechanisms are available for such scheduling, an appropriate one will be selected for supporting scheduling in MAI^{2}L.

Execution. The execution portion runs the procedure defined by the scheduled task. A monitoring process must evaluate the execution, to take unexpected events into account. It will be supported by the situation assessment component which allows for a sophisticated execution monitoring that goes beyond simply detecting expected effects of actions.

We have provided a summary of the conceptual foundation for MAI^{2}L, as well as the requirements for the MAI^{2}L ontology and run-time environments. The implementation of MAI^{2}L allows for construction of multi-agent systems in a wide variety of application domains.

References

1. J. F. Allen. Towards a general theory of action and time. *Artificial Intelligence*, 23:123–154, 1984.
2. Alastair Burt, editor. *A Preliminary Agent Model: Agent Communication.* Deliverable D-II.2.ii. IMAGINE, Esprit Project 5362, August 1991.
3. Alastair Burt. Plan generation in MAI^{2}L. Forthcoming, 1994.
4. M. P. Georgeff and A. L. Lansky (eds.). *Reasoning about Actions and Plans.* Morgan Kaufman, 1987.
5. L. P. Kaelbing and S. J. Rosenschein. Action and planning in embedded agents. In P. Maes, editor, *Designing Autonomous Agents.* MIT Press, 1991.
6. Michael Kolb and Donald Steiner. MAI^{2}L, a multi-agent implementation and interaction language. Forthcoming, 1994.
7. Donald Steiner, Alastair Burt, Michael Kolb, and Christelle Lerin. The conceptual framework for MAI^{2}L: An overview. Internal Report IR-II.2.xvii, IMAGINE, Esprit Project 5362, Siemens, September 1992.
8. J. R. Searle. *Speech Acts.* Cambridge University Press, Cambridge, 1969.
9. Y. Shoham. Agent-oriented programming. Technical Report STAN-CS-1335-90, Comp. Sci., Stanford Uni., 1990.

10. Donald Steiner, Dirk Mahling, and Hans Haugeneder. Human Computer Cooperative Work. In M. Huhns, editor, *Proc. of the 10th International Workshop on Distributed Artificial Intelligence*. MCC Technical Report Nr. ACT-AI-355-90, 1990.
11. Donald Steiner, editor. *A Preliminary Agent Model: Cooperating Agents*. Deliverable D-II.2.i. IMAGINE, Esprit Project 5362, Siemens, August 1991.
12. T. Winograd and F. Flores. *Understanding Computers and Cognition: A New Foundation for Design*. Ablex Publishing Corp., Norwood, New Jersey, 1986.

Designing good pursuit problems as testbeds for distributed AI: a novel application of genetic algorithms

Mauro Manela and J. A. Campbell
Department of Computer Science
University College London
Gower Street
London WC1E 6BT
England

e-mail: {mmanela, jac}@cs.ucl.ac.uk

Abstract

A basic N x M instance (game) of the Pursuit Problem is one in which N pursuing agents try to capture as many as possible of M prey agents by surrounding them, on a rectilinear grid. The 4 x 1 game has been considered as a testbed for comparing the effectiveness of different multi-agent distributed architectures, and the 6 x 2 game has received a little attention. This paper reports a systematic exercise in evaluating the quality of pursuit games as potential testbeds for distributed artificial intelligence (DAI). Genetic algorithms (GAs) have been used both to optimise low-level architectural features of agents and to search the (N, M) space of games. The conclusion from experiments is that $(M + 4) \times M$ games have the right complexity to be good testbeds, provided that $M > 4$. Additionally, the paper demonstrates the usefulness of GAs as tools to help DAI designers, and argues that boredom is a concept that deserves consideration as a feature of general agent architectures.

1 Introduction

Research in Distributed Artificial Intelligence (DAI) often considers the problem in which a group of independent problem-solvers (agents) needs to accomplish a given goal that is beyond the capability of any agent alone. This requires cooperation between the problem-solvers since one agent cannot simply proceed to perform its own actions without considering what the other agents are doing ([11]). Consequently, several alternative approaches to the distribution of knowledge and control among cooperative problem-solvers have been proposed (see [9] and [6] for examples).

As argued in [4], these approaches are usually still too specialised and project-specific. In other words, it is not clear whether a DAI architectural idea developed and shown to be effective for a given project is *generally* good and useful, because of the difficulty of comparing results obtained in different environments. This has probably been one of the motivations for the common choice, by more than one research group, of the Pursuit Problem as a reference problem for evaluating alternative approaches to distributed reasoning and studies of their performance. We give further information about the history of this problem in DAI in the next section. Here, we simply introduce the problem as first proposed by Benda, Jagannathan and Dodhiawala [3].

In this problem, two classes of agents (pursuers and pursued, or blue and red respectively) move on a rectilinear grid. Blue agents have the goal of blocking the path of any red agent. We describe a problem with N blue agents and M red agents as an $N \times M$ game. In most published studies of the Pursuit Problem, $N = 4$ and $M = 1$ (where the criterion for success for the blue team is simple: its agents capture the red agent by surrounding it before it can leave the grid by crossing a boundary). There have been some informal mentions of the 6×2 game.

For any problem that is proposed as a general DAI testbed, some clear measure or measures for how well the problem is solved by the given set of cooperating agents must exist. Now consider the architecture of the individual agents and the cooperative that they form. This architecture will have both "low-level" and "high-level" components, where the former are present in basically the same manner in any reasonable architecture. In a sense, they make up a minimal set of the features that an agent needs in order to have any claim to be "agent-like". The choice of high-level components in an architecture allows DAI researchers to be creative in their designs for effective agents and collectives. In the present state of DAI, the question of what high-level components to select is wide open - which is why the issue of testbeds and test problems for comparison of different architectural approaches is currently alive.

Against this background, we can state a simple but desirable property for test problems. Suppose that any measure of success has the value 0 for total failure and 1 for a perfect result. Then, with low-level components alone in place, a group of agents should be capable of a performance that does not rate 0 by the given measure(s) of success (because the high-level components would be handicapped by an unknown amount through having to remedy deficiencies in the low-level structures in addition to contributing to successful overall problem-solving performance). On the other hand, the low-level components alone should not lead to performance that has a rating too close to 1, or the improvements that are left for high-level components to make would not be large enough to put these components to a serious test.

On the informal evidence, the 4×1 pursuit game is not demanding enough to be a good testbed by these criteria. Published reports of experience with the 6×2 game are not yet conclusive. However, if the experience is eventually unsatisfactory, this is not enough to disqualify the general $N \times M$ Pursuit Problem from consideration as a good DAI testbed. Obviously the (N, M) space would then need further exploration. Moreover, the low-level characteristics of the agent architecture would add further real or notional parameters, increasing the dimensionality of the space that one would have to search in order to find pursuit games that could be good DAI testbeds. The size of the search might then discourage further investigation of the Pursuit Problem as a topic of any value for DAI.

The first purpose of our paper is to offer a remedy for this difficulty in advance, and to report on the results (including good pursuit games) that we have found. The difficulty is basically one of multidimensional optimisation, for which a cure in many instances is the use of genetic algorithms (GAs). We have found that a GA approach is productive here also.

Our second contribution to DAI is in suggesting a concept that can play a useful part in agent architectures: *boredom*. We have arrived at this suggestion partly by considering what architectural aspects of agents and collectives can be parametrised to allow the use of GAs, and how to parametrise them.

The organisation of the paper is as follows. In the first sections below we motivate and introduce the necessary concepts and tools. This is followed by showing the results of the simulations on the 4×1 game and the straightforward application of agents with the optimised architecture for the 4×1 game on games involving more agents and prey. We then develop a slight elaboration of the basic architecture with the introduction of *boredom*. Using the capabilities provided by the GA, we search for good combinations of parameters in the extended architecture and show that it provides better performance than for corresponding situations where boredom is not present. Finally we present results of the best values found by the GA in a class of difficult games, and hence suggest games that have just enough complexity and structure to be of interest when comparing alternative architectures.

2 Pursuit Games

The Pursuit Problem has been a popular testbed problem for evaluating the performance of alternative approaches to distributed reasoning since it was first proposed in [3]. It models a configuration of two classes of agents, *red* and *blue*, which move on a rectilinear grid. According to [9] the Pursuit Problem as formulated by Benda considers only four blue agents, positioned at different locations on a grid, and whose task is to capture a fifth red agent by surrounding it, i.e., a 4×1 game.

During a move in the game, any agent may remain at its current position or move by one unit in either a horizontal or vertical direction. Red agents move randomly in any direction that is not blocked by a blue agent. No two agents, regardless of type, may occupy the same location simultaneously. The movement of any red agent is limited to its current location or any of the unoccupied capture positions surrounding it. A red agent moves at random among these choices, and is considered captured if all its directions of movement are blocked by blue agents.

Gasser *et al.* [5] have explored the 4×1 game using a coordination framework based on patterns of settled and unsettled problems. Stephens and Merx [10] have compared particular versions of 3 different control strategies (local control, distributed control, and central control) and have arrived at conclusions similar to those of Benda *et al.* [3], i.e. that the central control strategy is superior to the other two although less efficient (and less robust, e.g. in terms of malfunctioning of the agents) than distributed control. Levy and Rosenschein ([9]) have approached the same game by incorporating global goals into the local interests of all agents through the use of techniques of game theory.

In a $N \times M$ game N blue agents, positioned at different locations on a grid, attempt to capture as many of the M red agents as possible. A blue agent never attempts to occupy a position that is occupied by a red agent. However, it may try to occupy a position that is occupied by another blue agent, in which case a conflict arises. Blue agents are assumed to have accurate knowledge of the position of all red agents and to have some sensing capability to detect that they are in conflict. Apart from this, no blue agent has any knowledge about any other blue agent's position. A red agent, once it leaves the grid, is assumed not to move further. Finally, all agents are assumed to move at the same speed.

3   An experimental framework

One of our basic purposes is to define a convincing minimal architecture of simple agents and study its performance on different sets of pursuit games. Within this framework agents should pursue their local goals only, and interactions between agents should be local and as limited as possible. Our second basic purpose is to examine the relative quality of different $N \times M$ games within the framework, to find the games that look best as testbeds for agent architectures completed by alternative high-level design features.

We have built a system of programs to implement pursuit games that follow the same general line as the one in Ref. [10]. The main difference is that we allow arbitrary numbers of agents instead of considering only the simplest 4×1 game. Another major difference between our implementation and the one reported in Ref. [10] is that we allow a blue agent to release a captured position that it has occupied, with the provision that it should communicate (by broadcasting) when it captures or releases a given position.

For ease of comparison with Stephens and Merx [10], each pursuit game was run on 30 initial grid configurations which were generated by placing the red agents symmetrically with respect to both the horizontal and vertical axes. Along each axis the minimum distance between neighbouring red agents was one unit square. Blue agents were placed randomly on the board subject to the condition that they should not be on the central horizontal and vertical lines. Figure 1 illustrates one such initial configuration of five red agents in a 10 × 10 grid.

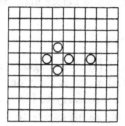

Figure 1: Initial configuration of five red agents

4 Performance metrics

Stephens and Merx have considered four performance metrics that summarise the performance of a given architecture in a 4 × 1 game: (1) pursuit outcome, (2) capture ratio, (3) success ratio, and (4) success efficiency. Pursuit outcome, success ratio and success efficiency will be discussed in sections 4.1, 4.2 and 4.3 respectively. *Capture ratio* refers to the number of experiments that result in a capture divided by the total number of experiments run. In a general pursuit game there is a high probability of capturing at least one prey per experiment, so this metric is not particularly helpful unless a good generalisation can be suggested.

The purpose of using a test problem may vary for different architectures. In particular, an architecture may be tailored to work in environments where there are stringent limits on communication bandwidth. Another may be designed to work in domains where conflicts occur very often and hence should be exercised in test problems with conflict-rich characteristics. We would expect these two types of situation to occur quite often in future applications, and to require at least one additional metric so that the architectures can be assessed adequately.

Even at a low level, architectures are often likely to contain primitive communication and conflict-resolution mechanisms. We therefore propose low-level indices that relate to them [1]. We consider the number of conflicts that occur

---

[1]In the architectures that we have examined, communication is very limited and occurs explicitly when an agent captures/releases a given capturing position. Conflicts refer to the situation when two blue agents want to occupy the same position.

per experiment and also the number of messages broadcast per experiment, as very simple measures of how much communication and conflict take place in each set of games. These are simple indices, but they offer hints of the intrinsic complexity of a given pursuit problem with respect to coordination of agents. The standing challenge is for a designer with particular high-level architectural ideas to see how those ideas, when implemented, can improve performance by comparison with the same test game before high-level treatment.

4.1 Pursuit Outcomes

A game is considered finished if: (i) all red agents are either captured or have crossed the grid boundaries, (ii) a preset limit on the number of iterations is reached. The latter condition is necessary to limit the number of iterations in one game to a reasonable value. In all games that we have played, we found that 100 iterations was a suitable value since the board configurations reached after that number of steps were fairly stable and apparently did not change much afterwards.

In the same way as for the 4×1 game considered in Ref. [10], once the termination of a particular game is reached any of the red agents can be in one of the following states: in a *capture* when it is completely surrounded by blue agents; in a *stalemate* when it is surrounded on three sides; in an *escape* when it has at most two captured positions occupied by blue agents. Also, a red agent will be in a *supercapture* state when it is surrounded by less than three blue agents but the remaining captured positions are blocked by other red agents or by blue agents that are surrounding a different red agent. In this way a supercapture reflects a more sophisticated capture by the blue agents, since red agents can be captured using less than four blue agents per red agent (in contrast with the 4×1 game). In fact, if it were not for this coordination, the blue agents could never capture both red agents in a 6×2 game.

4.2 Success Ratio

The *success ratio* as defined by Stephens and Merx is the expected value of a weighting function $f(.)$ that maps the outcomes to values in the closed interval $[0, 1]$. The values chosen for $f(.)$ were (following [10]) 1.0, 0.5, and 0.0 for capture, stalemate and escape, respectively. We give a supercapture the rating of 1.5. Since we consider games with more than one red agent, success ratio here is the ratio of the above quantity to the number of red agents.

In addition, we have considered a *normalised success ratio* which is the ratio of the success ratio to its maximum achievable value. For example, in a 6×2 game the latter is equal to 1.5 since red agents can always be supercaptured in principle. On the other hand, in a 5×7 game the maximum allowed value for the success ratio is 1.0/7 since in each experiment at most one prey can be captured. A common measure across such disparate games is therefore needed.

4.3 Success Efficiency

Stephens and Merx define *success efficiency* as an efficiency metric that can give some basis for comparing the performance of different architectures. The calculation of this metric depends on a global assignment of agents to capture positions. In games involving more than five agents, this procedure may lead to difficult and time-consuming computations. We have therefore defined success efficiency in the following way. For each prey captured (or supercaptured), we compute the ratio of the Manhattan distance travelled by the prey to the average Manhattan distance traversed by all blue agents having *captured* or *blocked* a given position on that prey. A prey that ends in a stalemate or escape has corresponding pursuit efficiencies of 0.5 and 0.0. The success efficiency with respect to that prey is then defined as

$$\frac{\sum_{i=1}^{n_{sc}} E_i f(sc) + \sum_{i=1}^{n_c} E_i f(c) + n_s f(s) + n_e f(e)}{n_{sc} + n_c + n_s + n_e}$$

where E_i is the capture efficiency with respect to the i-th red agent that has been either captured or supercaptured; n_{sc}, n_c, n_s, and n_e are the number of red agents that become involved in a supercapture, capture, stalemate or escape. The success efficiency of the entire architecture is the average of the above measure over all prey.

5 Autonomous Agents, Evolution and Genetic Algorithms

There are two separate issues in using evolutionary metaphors for the design of autonomous agents. The first is based on the experimental evidence that through aeons of evolutionary time, nature has developed autonomous organisms on Earth that rely in varying degrees on *sociability* for their survival. This offers some motivation for the generation of algorithms or machine learning techniques such as Classifier Systems [7] that evolve syntactically simple rules to guide an agent's performance in an arbitrary environment. The identification of what architectural issues are likely to be at the foundations of the design of autonomous agents may profit by inspiration from how natural organisms survive and cooperate among themselves.

The second issue is how to use algorithms based on evolution so as to optimise the design of autonomous agents for a specific task. This is the approach taken in the present paper. One such type of algorithm, a *Genetic Algorithm* (GA), has been developed by Holland [8]. GAs are based on the mechanics of genetic variation and natural selection. They have become increasingly popular in recent years due to their simplicity and elegance and to their power to discover good solutions rapidly for difficult high-dimensional problems.

In the simplest form of the GA, problem parameters are encoded in bit strings which play the role of artificial chromosomes, while individual bits play the role of genes. Each *individual* therefore comprises a candidate solution to a specific problem and has associated with it a corresponding *fitness*, i.e. a nonnegative figure of merit that measures the adequacy of the related solution. Individuals with high fitness values have a higher probability of being selected for producing multiple copies of higher-fitness individuals for the next *generation*. The *selection* of high-fitness individuals is combined with genetic operators such as mutation (flipping individual bits) and crossover (exchanging substrings of two parents to obtain two offspring), all applied in a probabilistic way to produce a new generation of individuals. The GA is judged to be successful if it can evolve a population of highly fit individuals as a result of iterating this procedure through successive generations. References [7, 8] contain detailed discussions of the basics of GAs.

Much as in nature, a given problem can be explored successfully by a variety of different classes of agents, each class with its own specific behavioural patterns. However, we would like to restrict the number of allowable behaviours to a bare minimum such that they remain non-problem-specific at least in the domain of Pursuit Games. This is for to the following reasons: (*i*) as argued by Anderson and Donath [1] in the domain of robot autonomy, the arbitrary addition of primitive behaviours may eventually serve no meaningful purpose; (*ii*) if the agents' number of behavioural responses is large, it may make the discrimination of the experimental results more complicated and hence may not contribute to better and more refined theories of social interaction; (*iii*) we would like to find behavioural patterns that emerge out of the interaction of a minimal number of primitive structures.

If the basic architecture can provide a reasonable performance for a particular problem, so much the better. If not, the system designer may fill the existing gaps with personal knowledge about the specific problem domain. At the same time, as argued above, a simple architecture functions as a baseline for assessing the possible difficulty a given problem may possess, and hence the suitability of trying to solve it with powerful algorithms on the one hand or with very simple approaches on the other. This, we believe, may offer a speed-up in the design cycle for the construction of suitable multi-agent architectures, e.g. in connection with the Pursuit Problem.

Next we discuss the process of design of a simple architecture that can be evolved through the application of a GA. Based on the remarks above, the necessary steps to be taken are: (*i*) the identification of basic dimensions in the agents' design, (*ii*) their parametrisation in a convenient way, and (*iii*) the definition of a suitable index or indices of merit for the performance of a given architecture in a set of games or tests. These points are explored in the next section.

6 Design issues

In the process of designing autonomous agents using a GA approach we were confronted with the following issues: (*i*) What questions should DAI designers ask? (*ii*) How can we apply GAs to the Pursuit Problem?

One answer to (*i*) deals with the problem of defining *which agent does what, when* [5]. With regard to (*ii*) we were faced with the problem of finding suitable dimensions of agent architectures to allow for convenient parametrisation and genetic search. It is obvious that some dimensional choices may be meaningless or not interesting while others may turn out to be useful not just because they promote the applicability of GA methods but also because they suggest new interpretations or ideas to the DAI designer. In trying to identify interesting dimensions for agent architectures (i.e., *what is parametrisable?*) we found ourselves asking the same question as (*i*) above, recast in the language of reactive architectures: *what are the observable quantities that should be measured and what are the thresholds for reactions? Are the thresholds based on correlations, predictions? And if so, of what and with what? What actions should be taken when a given threshold is exceeded?*

In trying to give an answer to (*ii*) we were drawn to the fact that in a multiagent system each individual agent should have at least a minimal capacity to plan its activities and coordinate them with those of other agents. It should also be provided with some minimal capacity to deal with conflict situations whenever they occur. In many problems, no agent is capable of performing a given task by itself. Hence agents should rely on coordination with others so that the community of agents will be able to accomplish the desired goal. In a complex pursuit game, it is not sufficient for an agent to block a given prey's direction of motion as there may not be enough other agents to block the prey's other possible directions for free movement (consider the case of a 5×7 pursuit game). As a consequence, if we want to endow even minimal simple agents with some capacity to cope with more complex games we have to make a choice about how they should react in these particular situations (subject to the fact that the agents have no global knowledge and very limited skills).

Bearing the above points in mind, the effect of our work (over a long period of study) has been to concentrate our design on three main issues: (*i*) prediction, (*ii*) conflict resolution and (*iii*) boredom.

Prediction refers to the agents' capacity to identify certain patterns/situations and immediately realise a possible version of their future consequences. In many circumstances agents should be designed with an innate capacity to react to unforeseen pattern/situations as this may enhance their efficiency (and sometimes their survival when confronted with dangerous environments).

Prediction also offers a finer control in the planning capability of an agent, as it allows the possibility of recognising future conflicts, or anticipating its target position (e.g. in a pursuit game a given agent chasing a prey could "cut corners" in its pursuit, if it were able to recognise correctly some regularity in a prey's moves). On doing so it could reduce its expenditures of energy and improve the efficiency of the whole community of problem-solvers.

Boredom is the recognition by an agent of certain patterns/situations that are not changing and that correlate with the agent's inability to accomplish its primary goal or to the community's inability to solve a given problem. An agent (even a very simple one) ought to perceive (or be told) that it is "not doing well" and it should have the ability (even if it is very limited) to look at the environment and reach the conclusion that something in its immediate neighbourhood should be changed to give any hope of improving its current state. Since our primary goal is to design an architecture that is very simple, we must include this issue as a low-level architectural choice if we include it at all. Fortunately a low-level version is realisable, using no more than the ideas we have just mentioned. Even when dealing with high-level design choices, however, a designer will be faced with similar issues.

Within the scope of the pursuit problem, conflicts occur because two agents may want to occupy the same position simultaneously. In our implementation we allowed conflicts to occur and be solved by the agents following very simple rules [2]. Agents were permitted one step of discrete time in which to resolve their conflicts; after this, any agent remaining in conflict over an intended move was required to stay in its current position.

6.1 Prediction

Since we wanted to keep the architecture as simple as possible, we decided to allow agents to predict only the prey's moves (and, for example, not to allow other inferences about a prey's internal state or structure). Bearing in mind the geometrical flavour of the pursuit problem, in which straight lines belong to the set of primitive concepts, the simplest predictive choice was to allow the agents to predict the prey's moves in a linear manner. Two different *decision modules* were implemented and both architectures were subjected to genetic search (which also included the search for their parameters). Figure 2 depicts one of these modules.

---

[2]e.g., in one implementation an agent engaged in conflict performed in a *polite way* by changing its intended move to another direction which was chosen (from a set of alternatives) in a random way or by taking its second-best alternative from the same set.

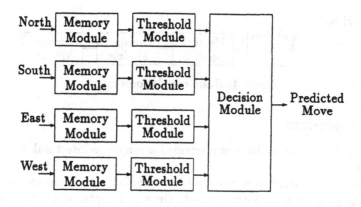

Figure 2: Basic predictor model

In this module the prey was considered to be moving in a given direction if the corresponding memory count was greater than or equal to the threshold. In the second option a direction was accepted as valid if the ratio of its corresponding memory count to the sum over all memory counters was greater than the threshold value. The decision module was included since in both implementations two different directions could be accepted as valid.

Once a predicted direction of move was found, the agent used a simple linear rule by which it projected the prey's future position on the direction of move. The final decision on where to move in the next step was made by comparing the distance towards the target position and the corresponding prediction. If the former was bigger, the agent moved to the latter. Otherwise it moved to the target position.

In the same way as above, we considered the design of memory modules throughout the entire architecture in the simplest way possible. This is illustrated in figure 3.

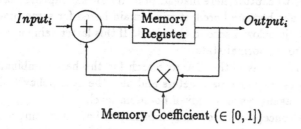

Figure 3: Memory model of past observations

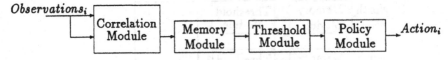

Figure 4: Basic correlation scheme

6.2 Boredom

Figure 4 shows how we have implemented a simple architectural correlation-based scheme.

Whenever an agent captures a position it broadcasts a message to indicate that that position is no longer vacant. Once in a captured position an agent will tend to remain there by following the prey's movements. If, however, insufficient other agents have turned up to surround the prey after some interval, the agent may become bored (i.e. when its associated boredom level exceeds a given threshold value) and eventually may release its position. The information that the prey has not been surrounded is correlated with information on the prey's movements, such as change of direction, detection of movement and the number of agents surrounding the prey. Once an agent releases a position it may for a number of iterations (i) stay in the same position, (ii) choose any random movement, (iii) backtrack, i.e., move in the direction opposed to its preferred choice of movement or (iv) when bored, choose the best alternative movement apart from following the prey and then target the nearest capture position that is available.

The last option has several interesting consequences. First, for a short period of time the bored agent is broadcasting a very primitive and "deceitful" message ("*I intend to release a captured position*") since some agent in the vicinity of the bored one may thus opt to target the available position. Because the bored agent may also try to target the released position in the next move, this will result in conflict. It also allows the bored agent to change its immediate target, since if it is blocking a capture position adjacent to another prey, it may decide to capture here instead of at its former capture position.

Once an agent has become bored, it will remain bored for some time until its boredom level falls below a second threshold. If this latter value is reached, the agent returns to its normal state.

For this module, we used the GA to search for the best combination of correlation strategies, memory parameters and also the best policy to be implemented once an agent reaches a given boredom level.

We believe the concept of boredom is a primitive but interesting architectural idea that is worth considering for inclusion in the design of many types of agents and DAI systems. In the pursuit problem (as in many other conceivable DAI applications) "a little boredom goes a long way". This is illustrated by Table 1, which depicts the best pursuit outcomes found by the GA on a 5×7 pursuit game for architectures including or excluding boredom (see Section 7 below for further details on the optimisation issues).

Boredom			No Boredom		
Escapes	Captures	Stalemates	Escapes	Captures	Stalemates
185	10	15	200	2	8

Table 1: 5×7 Pursuit outcome

The introduction of boredom allowed captures to occur in one-third of the games while stalemates occurred in half of them [3]. This is reflected by the normalised success ratio, which was equal to 58% in the architecture with boredom as opposed to only 22% in the simple architecture.

6.3 Conflict Resolution

To examine conflict resolution, we implemented several simple policies and used the capability of the GA to search for the most effective one. In a pursuit game involving more than 4 pursuing agents the simplest form of conflict is related to the fact that two or more agents may want to occupy the same location simultaneously. In accordance with the design restriction of having no global knowledge, agents resolved their conflicts by changing their choices of where to move next following very simple rules. If an agent persisted in conflict after trying to follow the rules, it was forced to stay in the same position. Conflicts also occurred due to the prey's movements since an agent that had captured a position and was following the prey's move could be dragged to the same position as another agent. Whenever this occurred, in our approach, the docked agent was forced to undock.

7 Optimisation Issues

We have so far described the model parameters that we have subjected to genetic search and have discussed some performance metrics that can be used for comparing the efficiency of different architectures. Recall that one of our aims is to suggest games that look best as testbeds, i.e., games in which the low-level components of an agent's architecture do not lead to a performance that has either a very low or a very high rating. As such, a game will be considered very complex if a community of simple problem-solvers cannot be found such that it attains a minimum level of performance with respect to the former. On the other hand, if we can find an architecture of simple agents that achieves a high performance in a given set of games, we shall consider

---

[3]In effect one out of the two captures in the no-boredom architecture occurred because a red agent was surrounded by three blue agents and another red agent that had previously moved to an effective "capture" position by being stationary just out of the grid boundaries (so it should be considered as a supercapture, although it was a very opportunistic one).

the set of games as not being of interest for the purposes of a DAI testbed. Stated in other terms, given a set of games we shall use a GA to search for an architecture that achieves the highest performance in it. A set of games will be considered as a good testbed if the performance of this architecture falls within a given range.

In this way, the search for good games will comprise three phases: (1) finding sets of games that look interesting in principle; (2) submitting each set of games as a problem to be solved by an architecture of simple agents that will be evolved using a GA; and (3) identifying among the set of games those for which the performance of the architectures evolved by the GA falls within a given range. In this section we shall consider how (2) can be accomplished.

Each candidate solution produced by the GA will comprise the parameters of a given agent's architecture, the performance of which will be calculated by requiring it to play a given set of games. A GA evolves sets of candidate solutions towards *populations* of *fitter* solutions. Within our interpretation, a solution (or an agent's architecture) will be considered satisfactory (or fit) if a corresponding community of agents (with the given parameters) achieves a reasonable performance when playing a given set of games. Ideally we would like to regard the GA as being successful in its search for the architecture's parameters if it can produce solutions near the global optimum, i.e., if no other set of parameters exists such that these result in a better performance. This brings into perspective the following interrelated points:

- What is the criterion to be optimised by the GA?

- What will be considered as a good "fit" solution?

- How does a given architecture that was found by optimisation against a fixed set of games perform with respect to different initial configurations?

The natural candidates for the optimisation criterion were initially the success ratio and the success efficiency. We first examined these criteria with respect to the 4×1 game and the performance metrics suggested in Ref. [10]. Once a particular architecture was found by the GA we inspected the solutions via an animated visual display. Based on this, we observed that solutions produced by optimisation of the success-efficiency metric were of worse quality than those produced by optimising the success-ratio criterion. The former led to communities of agents that hovered in the grid without approaching the prey in a consistent way, and yet achieved high values of the efficiency metric.

The last observation can be understood by the fact that, if a blue agent does not move but still manages to block one direction of the red agent, it is behaving in a very efficient way. The GA managed to find architectures where agents seemed to be exploiting this approach passively rather than actually exploring the possibilities that were open to them. We have observed that minimisation of the sum of the success-ratio and success-efficiency metrics resulted in systems with the most effective game-playing behaviour. Other metrics could be generated by different weighted combination of the performance metrics. However, we have used the success ratio instead since it produced good-quality solutions and it was simpler and faster to calculate.

Since theoretical information about the global optimum was in general unavailable, we decided to use the GA's results as natural candidates. In the 4×1 game this offered no problem as solutions achieving the maximum allowed value were always produced. In more complex games, however, we took the viewpoint that (lacking any other significant information) the solutions found by the GA could be used as a baseline for comparison. The GA started with a population of 50 individuals chosen at random and we let it run for 3500 trials. The individual solution with the highest success ratio was chosen as the fittest one. In general, due to limits on our computational resources, we ran the GA once for each set of games. It is a known fact in the GA literature [7] that a GA can become trapped on local optima and hence running the algorithm several times is usually recommended when one can afford to do so. As we used it in very complex games that demanded a great number of computations, running the algorithm several times would have been prohibitive (although playing the set of games in parallel on several machines would make the task much easier).

Tests were performed to assess the variance of the solutions generated by the GA using 7×3 as an example. We ran the GA 10 times and collected statistics concerning the results. The standard deviation of the best success ratio obtained in each run was more than an order of magnitude smaller than the corresponding mean. We observed that in all experiments conducted, the GA managed to find good-quality solutions and did not seem to get trapped prematurely in any particular area of the search space. This is particularly pleasing once it is recognised that the GA was attempting to find good combinations of very different parameters such as thresholds and policies to be adopted when a given threshold was achieved. The solutions provided by the GA continued to perform well when we considered different sets of initial configurations.

8 Experimental Results

We have conducted experiments to provide suggestions of answers to the following questions:

- Is the 4×1 game sufficiently complex to be useful as a testbed for different architectures?

- How does a particular architecture that was optimised for a 4×1 game behave in more complex games?

- In case an answer to the first question happens to be negative, what games have enough structure and complexity to be exploited as useful testbeds?

The last question is particularly important, and our response to it constitutes a significant novel result for this paper.

The answer to the first question proved to be particularly easy: a 4 × 1 game could be played successfully by a community of very simple agents with a performance equal to that of the best architectures reported elsewhere. Table 2 illustrates the performance of an architecture in which agents relied only on their rudimentary predictive capabilities and conflict-resolution strategies obtained through optimisation via the GA (in this table, Conflicts refers to the number of conflicts per experiment while Messages refers to the number of capture/release messages per experiment). Since the number of agents in a given conflict was almost always two, the number of actual messages broadcast was less than 15 per experiment. This compares rather well with the results reported in [10].

Escapes	Captures	Stalemates	Success Ratio	Conflicts	Messages
0	29	1	98.33 %	3.067	7.533

Table 2: Performance metrics of predictive architecture in a 4 × 1 game

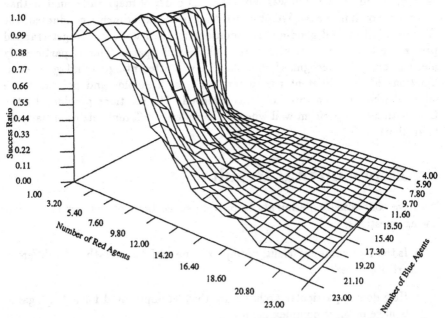

Figure 5: Success-Ratio performance of an optimised "4 × 1" architecture in complex games

The inclusion of boredom in the basic architecture resulted in 100% success ratio with a very much reduced number of conflicts per experiment (1.5) and a slightly higher number of capture/release messages per experiment (8.1). Hence, 100% success ratio can be achieved in a 4×1 game with a minimal amount of coordination and communication.

Figure 5 illustrates the success-ratio performance of a simple architecture of agents optimised by the GA when used to play more complex games. It is consistent with the observation [10] that a distributed-control system degrades gracefully and that enough simple agents can be added to achieve a desired success ratio.

Figure 5 suggests degradation of the performance metric in a Gaussian way where the standard deviation is proportional to the number of blue agents. This can be seen from Figure 6 where we plot success ratio against the ratio of red to blue agents. Figure 5 also serves to suggest areas of the game space where games of greater complexity (i.e. in which the performance of the agents' architecture achieves very low values) can be found. It must be emphasised, however, that it reflects somehow the performance of an architecture specially optimised for a 4×1 game when confronted with more complex games. (A specific architectural optimisation for each particular game would in principle produce better results).

Figure 6 shows a sharp transition in the performance metric around the value 0.3 for the ratio of red to blue agents. The "knee" of the curves is close to the value 0.6 and the curves flatten for values greater than or in the vicinity of 1.2. Using this information as a subjective guideline, we reasoned that games within the region [0.3, 1.2] formed an initial set of good testbed problems. Yet this still left us with a huge number of games to be evaluated. The next step was to narrow down the search to a class of games for which all

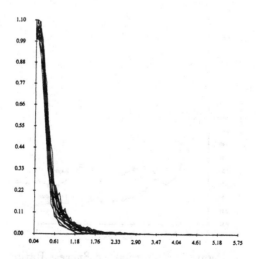

Figure 6: Success-Ratio performance against ratio of red to blue agents

the red agents could be captured (at least in principle) using all the available blue agents (albeit probably demanding a lot of coordination among them). Our interpretation of the data is that *the best class is the class of* $(M+4) \times M$ *games*. (For example, $N = M + 4$ holds on a line that is a good subjective separator of the high-gradient region and the plain-like region on the surface in figure 5). Incidentally, the 6×2 game is the first element of this class of games, for which the above ratio is exactly $1/3$.

Figure 7 shows the optimum success ratio found by the GA when used to search for the best value of the performance metric in this set of complex games. It illustrates the utility of implementing boredom in an agent architecture over one that does not take it into account. In this class of games the degradation in performance is exponential at first but falls off very slowly for games involving more than 7 red agents with the curves oscillating slightly but showing a steady and slow decline as the number of red agents increases. It appears to indicate that the increase in the number of agents above a suitable non-small value does not introduce much extra complexity into the games played. Values of M below or equal to 4 seem to lead to games that can be tackled with reasonable success by communities of very simple problem-solvers and as such should be playable with very high efficiency by any agent architecture with good high-level design features. Therefore it is possible to conclude that the 6×2, 7×3 and 8×4 games represent qualifying exercises for any such candidate architecture, while the serious tests for the qualifiers start with 9×5.

Figure 8 shows the number of conflicts occurring per experiment while figure 9 shows the number of capture/release messages broadcast. In all experiments conducted, almost all conflicts involved only two agents. Obviously an increase in communication should be taken into account in those archi-

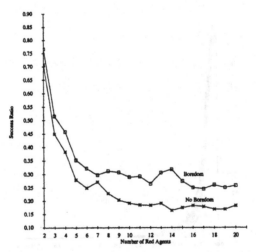

Figure 7: Unnormalised Success-Ratio

tectures where a conflict-resolution policy relies on agents exchanging some form of communication. But figure 9 in particular indicates that even a simple change in the features of agent design can produce a surprisingly large increase in the demand for communication bandwidth.

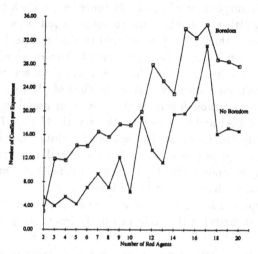

Figure 8: Conflicts per experiment

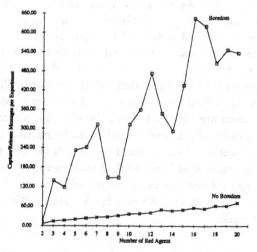

Figure 9: Broadcast messages per experiment

9 Conclusions

In this paper we have examined the design of autonomous agents using a GA approach to explore the space of pursuit games and to suggest games in this space that may serve as good testbeds for alternative DAI architectures. Based on the experimental evidence collected, we have concluded that the 4×1 game does not serve the purpose for which it was conceived, since a minimal agent's architecture using little coordination and communication can solve the problem so well that there is nothing left for more sophisticated architectures to improve upon. Our exploration indicates that the Pursuit Problem should not be ruled out as a good DAI testbed provided that the right games are considered. The experiments suggest quite strongly that in the class of $(M + 4) \times M$ games the simplest ones that have enough complexity to serve as useful supports for tests and evaluations of DAI architectures are those with $M \in [5, 7]$. For $M > 7$, the games certainly require more than low-level architectures if the pursuing agents are to achieve good performance, but the overall computational cost of playing them may be undesirably high when one considers that the outcomes may give little extra enlightenment to DAI designers.

The results reported here have a value beyond the domain of Pursuit Problems. We have demonstrated that DAI can profit from the application of GAs as experimental tools to assist the *design* of different architectures. The assistance is of two kinds. First, it allows searches of suitable spaces (here, both "architectural spaces" and the (N, M) game space) for good design choices, when no other approach can guarantee to make such choices. Second, it is an aid to thinking about the dimensions of the spaces, as one has to consider the question of what can be parametrised when one is trying to set up a problem for GA analysis. One may not know in advance which kind of assistance will be of greater value for a given problem. In the research reported here, we have been able to take advantage of both kinds of assistance. Most of the paper is devoted to discussion of optimal choices of low-level architectures and pursuit games. However, we have also identified some basic dimensions in agents' architecture that are potentially valid for many different DAI applications. Within this activity, we believe that the issues introduced through the concept of *boredom* are the most novel. In addition, although we do not have the space to discuss them here, we have also found that the concepts of thresholds and attenuation of communication, and "superagents" (assemblies of blue or pursuing agents that have reached positions relative to each other, e.g. while docked against red agents as mentioned in section 4.1, that it is in their best interests to maintain), are worth further study. We intend to report on them in future work.

Acknowledgement

Mauro Manela is sponsored by CAPES (Coordenação de Aperfeiçoamento de Pessoal de Nível Superior) - Brazil.

References

[1] T.L. Anderson and M. Donath (1990). Animal behavior as a paradigm for developing robot autonomy. In P. Maes (ed.), *Designing autonomous agents: theory and practice from biology to engineering and back*, 145-168. Cambridge, Massachussets: MIT Press.

[2] N. M. Avouris and L. Gasser (1992). *Distributed Artificial Intelligence: Theory and Praxis*, 1-7, Dordrecht: Kluwer Academic Publishers.

[3] M. Benda, V. Jagannathan, and R. Dodhiawala. On optimal cooperation of knowledge sources. Technical Report BCS-G2010-28, Boeing AI Center, Boeing Computer Services, Bellevue, Washington, August 1985. Cited in L. Gasser, N. F. Rouquette, R. W. Hill, and J. Lieb, (1989). Representing and Using Organizational Knowledge in Distributed AI Systems. In L. Gasser and M. N. Huhns (eds.), *Distributed Artificial Intelligence*, 2: 55-78, London: Pitman.

[4] L. Gasser and M. N. Huhns, Themes in Distributed Artificial Intelligence Research (1989). In L. Gasser and M. N. Huhns (eds.), *Distributed Artificial Intelligence*, 2: *vii-xv*, London: Pitman.

[5] L. Gasser, N. F. Rouquette, R. W. Hill, and J. Lieb, Representing and Using Organizational Knowledge in Distributed AI Systems. In L. Gasser and M. N. Huhns (eds.), *Distributed Artificial Intelligence*, 2: 55-78, London: Pitman.

[6] L. Gasser (1992). An Overview of DAI. In N. M. Avouris and L. Gasser (eds.), *Distributed Artificial Intelligence: Theory and Praxis*, 9-30. Dordrecht: Kluwer Academic Publishers.

[7] D. E. Goldberg (1989). *Genetic Algorithms in Search, Optimization and Machine Learning*. Reading, Massachussets: Addison Wesley.

[8] J. H. Holland (1975). *Adaptation in natural and artificial systems*. Ann Arbor, Michigan: University of Michigan Press.

[9] R. Levy and J. S. Rosenschein (1992). A Game Theoretic Approach to Distributed Artificial Intelligence and the Pursuit Problem. In Y. Demazeau and E. Werner (eds.), *Decentralized Artificial Intelligence III*, 129-146, Amsterdam: Elsevier Science Publishers B.V./North-Holland.

[10] L. M. Stephens and M. B. Merx (1990). The Effect of Agent Control Strategy on the Performance of a DAI Pursuit Problem. In *Proceedings of the 10th. International Workshop on Distributed Artificial Intelligence*, Bandera, Texas, October 1990.

[11] E. Werner, Cooperative Agents: A Unified Theory of Communication and Social Structure. In L. Gasser and M. N. Huhns (eds.), *Distributed Artificial Intelligence*, **2**: 3-36, London: Pitman.

Lecture Notes in Artificial Intelligence (LNAI)

Lecture Notes in Computer Science